Titles in This Series

Volume

1 **Markov random fields and their applications,** Ross Kindermann and J. Laurie Snell

2 **Proceedings of the conference on integration, topology, and geometry in linear spaces,** William H. Graves, Editor

3 **The closed graph and P-closed graph properties in general topology,** T. R. Hamlett and L. L. Herrington

4 **Problems of elastic stability and vibrations,** Vadim Komkov, Editor

5 **Rational constructions of modules for simple Lie algebras,** George B. Seligman

6 **Umbral calculus and Hopf algebras,** Robert Morris, Editor

7 **Complex contour integral representation of cardinal spline functions,** Walter Schempp

8 **Ordered fields and real algebraic geometry,** D. W. Dubois and T. Recio, Editors

9 **Papers in algebra, analysis and statistics,** R. Lidl, Editor

10 **Operator algebras and K-theory,** Ronald G. Douglas and Claude Schochet, Editors

11 **Plane ellipticity and related problems,** Robert P. Gilbert, Editor

12 **Symposium on algebraic topology in honor of José Adem,** Samuel Gitler, Editor

13 **Algebraists' homage: Papers in ring theory and related topics,** S. A. Amitsur, D. J. Saltman, and G. B. Seligman, Editors

14 **Lectures on Nielsen fixed point theory,** Boju Jiang

15 **Advanced analytic number theory. Part I: Ramification theoretic methods,** Carlos J. Moreno

16 **Complex representations of GL(2, K) for finite fields K,** Ilya Piatetski-Shapiro

17 **Nonlinear partial differential equations,** Joel A. Smoller, Editor

18 **Fixed points and nonexpansive mappings,** Robert C. Sine, Editor

19 **Proceedings of the Northwestern homotopy theory conference,** Haynes R. Miller and Stewart B. Priddy, Editors

20 **Low dimensional topology,** Samuel J. Lomonaco, Jr., Editor

21 **Topological methods in nonlinear functional analysis,** S. P. Singh, S. Thomeier, and B. Watson, Editors

22 **Factorizations of $b^n \pm 1$, $b = 2, 3, 5, 6, 7, 10, 11, 12$ up to high powers,** John Brillhart, D. H. Lehmer, J. L. Selfridge, Bryant Tuckerman, and S. S. Wagstaff, Jr.

23 **Chapter 9 of Ramanujan's second notebook—Infinite series identities, transformations, and evaluations,** Bruce C. Berndt and Padmini T. Joshi

24 **Central extensions, Galois groups, and ideal class groups of number fields,** A. Fröhlich

25 **Value distribution theory and its applications,** Chung-Chun Yang, Editor

26 **Conference in modern analysis and probability,** Richard Beals, Anatole Beck, Alexandra Bellow, and Arshag Hajian, Editors

27 **Microlocal analysis,** M. Salah Baouendi, Richard Beals, and Linda Preiss Rothschild, Editors

28 **Fluids and plasmas: geometry and dynamics,** Jerrold E. Marsden, Editor

29 **Automated theorem proving,** W. W. Bledsoe and Donald Loveland, Editors

30 **Mathematical applications of category theory,** J. W. Gray, Editor

31 **Axiomatic set theory,** James E. Baumgartner, Donald A. Martin, and Saharon Shelah, Editors

32 **Proceedings of the conference on Banach algebras and several complex variables,** F. Greenleaf and D. Gulick, Editors

33 **Contributions to group theory,** Kenneth I. Appel, John G. Ratcliffe, and Paul E. Schupp, Editors

34 **Combinatorics and algebra,** Curtis Greene, Editor

35 **Four-manifold theory,** Cameron Gordon and Robion Kirby, Editors

36 **Group actions on manifolds,** Reinhard Schultz, Editor

37 **Conference on algebraic topology in honor of Peter Hilton,** Renzo Piccinini and Denis Sjerve, Editors

Titles in This Series

Volume

Free Group Rings

CONTEMPORARY MATHEMATICS

Volume 66

Free Group Rings

Narain Gupta, Editor

AMERICAN MATHEMATICAL SOCIETY
Providence · Rhode Island

MATH
Sep/ae

1980 *Mathematics Subject Classification* (1985 *Revision*). Primary 20F05, 20F10, 20F12, 20F14, 20F26, 20C05, 20C07, 16A27, 20H25.

Library of Congress Cataloging-in-Publication Data

Gupta, Narain.

Free group rings.

(Contemporary mathematics, ISSN 0271-4132; v. 66)
Bibliography: p.
1. Group rings. I. Title. II. Series: Contemporary mathematics (American Mathematical Society); v. 66.

QA171.G88 1987 512′.4 87-12427
ISBN 0-8218-5072-5 (alk. paper)

SD
9/2/87

Table of Contents

TABLE OF CONTENTS

Preface

This MONOGRAPH deals with some aspects of *linear techniques* in Combinatorial Group Theory which have their origin in the work of Wilhelm Magnus in the thirties. The central theme is the identification and properties of those subgroups of free groups which are induced by certain ideals of the *free group rings*, the integral group rings of free groups. The subject has been developed extensively in the literature and my intention is to give a systematic and comprehensive account of some of its developments in a contemporary style. The choice of topics presented and their respective treatment is inevitably influenced by my own research over a long period. A brief outline of the topics dealt with is as follows.

CHAPTER I is introductory. Here we begin with a faithful 2×2 matrix representation of the quotient ring $\mathbb{Z}F/\mathfrak{x}\mathfrak{f}$, where $\mathfrak{f} = \mathbb{Z}F(F-1)$ is the augmentation ideal of the free group ring $\mathbb{Z}F$ and $\mathfrak{x}$ is any ideal of $\mathbb{Z}F$ contained in $\mathfrak{f}$. With $\mathfrak{x} = \mathfrak{r} = \mathbb{Z}F(R-1)$, R normal in F, we prove Schumann's result that $F \cap (1+\mathfrak{r}\mathfrak{f}) = R'$, the commutator subgroup of R. From this we deduce the well-known Magnus embedding of F/R' into a suitable ring of 2×2 matrices. After a brief introduction to Fox's Free Differential Calculus we formulate the so-called FOX PROBLEM: *Identify the subgroups of F induced by the ideals $\mathfrak{r}\mathfrak{f}^n$ for $n \geq 1$.* In the convenient language of free groups rings, we also formulate the well-known DIMENSION SUBGROUP PROBLEM: *Identify the subgroups of F induced by the ideals $\mathfrak{r}+\mathfrak{f}^n$ for $n \geq 1$.* We conclude the chapter with a self-contained new proof of the Fundamental Theorem of Free Group Rings due to Magnus, namely, the *ideal $\mathfrak{f}^n$ induces the n-th term of the lower central series of* F *for each $n \geq 1$.*

CHAPTER II is designed to illustrate the fact that the Magnus embedding provides a powerful tool in group theory. In particular, we derive some of the well-known results about F/R' including the solution of the conjugacy problem of F/R' due to Remeslennikov & Sokolov and Passi's theorem about the faithfulness of R/R' as a $\mathbb{Z}(F/R)$-module.

In CHAPTER III we give a detailed account of the Fox Problem including its recent solutions due to Gupta and Yunus. We also include the structure of the Fox modules and their fixed submodules due to Gupta & Passi.

CHAPTER IV contains an up-to-date development of the Dimension Subgroup Problem. Simplifications due to Hartley, Cliff & Hartley and Gupta yield an easy access to Sjogren's Fundamental Theorem from which most of the earlier results are deduced as corollaries. We also include an accessible account of Rips' counter-example to the Dimension Subgroup Conjecture. The remainder of the chapter is devoted to the dimension subgroups of metabelian groups. Here we include a theorem of Gupta, Hales & Passi about the dimension subgroups of finitely generated metabelian groups. We also include Tahara's description of the fourth dimension subgroups and conclude the chapter with a new result about the dimension subgroups of metabelian p-groups.

In CHAPTER V we describe a faithful matrix representation of the quotient groups $F/F \cap (1+\underline{r}_1 \cdots \underline{r}_n)$, $n \geq 2$, $\underline{r}_i = \mathbb{Z}F(R_i-1)$, R_i normal in F, due to Lewin and Yunus. We derive some important results about the residual nilpotence of these groups due to Yunus and C.K. Gupta & Passi. We point out some recent developments in the areas of generalized embeddings due to Kuz'min and Stöhr, and conclude the chapter with some general open problems in the area of free group rings.

This monograph is aimed at graduate students and other researchers in the area of combinatorial group theory. In particular, we assume some familiarity with the general terminology of free groups and group rings. Lack of my own expertise in the areas of <u>homological</u> and <u>geometrical</u> aspects of free group rings has prevented me from introducing these developments into the text. For these topics the reader is referred to the Lecture Notes by Karl Gruenberg (1970) and Joan Birman (1974). For other aspects of Group Rings the reader is referred, in particular, to Passman (1977), Sehgal (1978) and Passi (1979).

The idea for writing a monograph on Free Group Rings was originally suggested by Inder Bir Passi in 1980 after I had given a series of lectures at the Centre for Advanced Study in Mathematics, Chandigarh. Much of the material presented in Chapters III and IV has been developed since that time. I thank Inder Bir Passi for his help in organizing and collaborating on various parts of the Monograph. I also thank C. Kanta Gupta, Brian Hartley and Roger Lyndon for their encouragement during this project. This work was completed while I held a Killam Research Fellowship from the Canada Council and a research grant from the Natural Sciences and Engineering Research Council of Canada.

Narain Gupta
Winnipeg, January 1987

Chapter I

Magnus Embeddings and Free Differential Calculus

This chapter is introductory. Here we develop the language of free group rings and formulate the well-known Fox problem and the dimension subgroup problem.

1. SOME FAITHFUL 2 × 2 MATRIX REPRESENTATIONS

Let F be a free group with basis X and $\mathbb{Z}F$ be its integral group ring. Let $\mathfrak{f} = \ker(\varepsilon\colon \mathbb{Z}F \to \mathbb{Z})$ be the augmentation ideal of $\mathbb{Z}F$ and $\mathfrak{x} \le \mathfrak{f}$ be a two-sided ideal of $\mathbb{Z}F$. We begin with a faithful matrix representation of $\mathbb{Z}F/\mathfrak{x}\mathfrak{f}$.

1.1 *NOTATION*. Let Ω be a free $(\mathbb{Z}F/\mathfrak{x}\,,\ \mathbb{Z}+\mathfrak{x}/\mathfrak{x})$-bimodule with basis $\{\lambda_x,\ x \in X\}$ and

$$M = \begin{bmatrix} \mathbb{Z}F/\mathfrak{x} & \Omega \\ 0 & \mathbb{Z}+\mathfrak{x}/\mathfrak{x} \end{bmatrix}$$

be the ring of 2×2 upper-triangular matrices over Ω. □

1.2 *NOTATION*. The map

$$\psi\colon x \to \begin{bmatrix} x+\mathfrak{x} & \lambda_x \\ 0 & 1+\mathfrak{x} \end{bmatrix},\quad x \in X,$$

extends to a homomorphism of F into the group of units of M and its linear extension

$$\psi^*\colon \mathbb{Z}F \to M$$

defines a ring homomorphism. □

1.3 *NOTATION*. We denote by $\alpha_{ij}(u)$, $u \in \mathbb{Z}F$, the ij-entry of the matrix $\psi^*(u)$, i.e.,

$$\psi^*(u) = \begin{bmatrix} \alpha_{11}(u) & \alpha_{12}(u) \\ 0 & \alpha_{22}(u) \end{bmatrix}. \qquad \square$$

Clearly, $\alpha_{11}(u) = 0$ if and only if $u \in \underline{\underline{x}}$ and $\alpha_{22}(u) = 0$ if and only if $u \in \underline{\underline{f}}$. We prove,

1.4 *THEOREM*. $\ker \psi^* = \underline{\underline{x}}\,\underline{\underline{f}}$.

Proof. With $u \in \underline{\underline{x}}$, $v \in \underline{\underline{f}}$, $\psi^*(u)$ and $\psi^*(v)$ are of the form $\begin{bmatrix} 0 & a \\ 0 & 0 \end{bmatrix}$ and $\begin{bmatrix} b & c \\ 0 & 0 \end{bmatrix}$ respectively. Thus $\psi^*(uv) = \psi^*(u)\psi^*(v) = 0$ and it follows that $\underline{\underline{x}}\,\underline{\underline{f}} \le \ker \psi^*$.

Conversely, let $u \in \ker \psi^*$. Then $\alpha_{22}(u) = 0$ implies $u \in \underline{\underline{f}}$. Thus we may write u as

$$u = \sum_x u_x(x-1),\ u_x \in \mathbb{Z}F,\ x \in X.$$

Now,

$$\begin{aligned} 0 &= \alpha_{12}(u) \\ &= \sum_x \alpha_{11}(u_x)\alpha_{12}(x-1) + \alpha_{12}(u_x)\alpha_{22}(x-1) \\ &= \sum_x \alpha_{11}(u_x)\alpha_{12}(x) \\ &= \sum_x \alpha_{11}(u_x)\lambda_x . \end{aligned}$$

Thus $\alpha_{11}(u_x) = 0$ and consequently $u_x \in \underline{\underline{x}}$ for each $x \in X$. It follows that $u \in \underline{\underline{x}}\,\underline{\underline{f}}$ as required. $\square$

1.5 *COROLLARY*. $F/F \cap (1+\underline{\underline{x}}\,\underline{\underline{f}})$ is isomorphic to the group generated by $\psi(x)$, $x \in X$, as given by 1.2.

Proof. Since $\ker \psi = F \cap (1+\ker \psi^*)$, the proof follows by Theorem 1.4. $\square$

A particularly interesting case of Corollary 1.5 occurs when $\underline{\underline{x}} = \underline{\underline{r}} = \ker(\theta\colon \mathbb{Z}F \to \mathbb{Z}(F/R))$ (i.e., $\underline{\underline{r}} = \mathbb{Z}F(R-1)$), $R \trianglelefteq F$. A well-known result of Schumann (1935) states that $F \cap (1+\underline{\underline{r}}\,\underline{\underline{f}}) = R'$, the commutator subgroup of R. This, together with Corollary 1.5, yields a faithful matrix representation of F/R' known as the Magnus embedding of F/R' into M. Schumman's Theorem is a particular case of the following theorem.

1.6 *THEOREM*. (Enright 1968). Let $R, S \trianglelefteq F$ and let $\underline{\mathfrak{r}} = \mathbb{Z}F(R-1)$, $\underline{\mathfrak{s}} = \mathbb{Z}F(S-1)$ be two-sided ideals of $\mathbb{Z}F$. Then $F \cap (1+\underline{\mathfrak{r}}\underline{\mathfrak{s}}) = [R \cap S, R \cap S]$.

Proof. For $z_1, z_2 \in R \cap S$, $z_i - 1 \in \underline{\mathfrak{r}} \cap \underline{\mathfrak{s}}$ and $[z_1, z_2]-1$ $= z_1^{-1}z_2^{-1}((z_1-1)(z_2-1) - (z_2-1)(z_1-1)) \in (\underline{\mathfrak{r}} \cap \underline{\mathfrak{s}})^2 \leq \underline{\mathfrak{r}}\underline{\mathfrak{s}}$. It follows that $[R \cap S, R \cap S] \leq F \cap (1+\underline{\mathfrak{r}}\underline{\mathfrak{s}})$.

For the reverse inclusion, we consider a matrix representation of $\mathbb{Z}F/\underline{\mathfrak{r}}\underline{\mathfrak{s}}$ as follows: Let $\hat{\Omega}$ be a free $(\mathbb{Z}(F/R), \mathbb{Z}(F/S))$-bimodule with basis $\{\lambda_x : x \in X\}$, and let

$$\hat{M} = \begin{bmatrix} \mathbb{Z}(F/R) & \hat{\Omega} \\ 0 & \mathbb{Z}(F/S) \end{bmatrix}$$

be a ring of 2×2 matrices. Then the map

$$\hat{\psi}: x \to \begin{bmatrix} xR & \lambda_x \\ 0 & xS \end{bmatrix}$$

extends to a homomorphism of F into the group of units of $\hat{M}$ and its linear extension $\hat{\psi}^*: \mathbb{Z}F \to \hat{M}$ is a ring homomorphism. For any $u \in \mathbb{Z}F$, we write

$$\hat{\psi}^*(u) = \begin{bmatrix} \beta_{11}(u) & \beta_{12}(u) \\ 0 & \beta_{22}(u) \end{bmatrix}.$$

It is readily seen that for $u \in \underline{\mathfrak{r}}$, $v \in \underline{\mathfrak{s}}$, $\hat{\psi}^*(uv) = \hat{\psi}^*(u)\hat{\psi}^*(v) = 0$ so that $\underline{\mathfrak{r}}\underline{\mathfrak{s}} \leq \ker \hat{\psi}^*$. This yields, in particular, $F \cap (1+\underline{\mathfrak{r}}\underline{\mathfrak{s}})$ $\leq F \cap (1+\ker \hat{\psi}^*) = \ker \hat{\psi}$. Thus it suffices to prove that $\ker \hat{\psi} \leq [R \cap S, R \cap S]$.

Indeed, let $w \in \ker \hat{\psi}$. Then $\beta_{11}(w) \in R$, $\beta_{22}(w) \in S$ imply $w \in R \cap S$. Let $w = x_1^{\varepsilon_1} \dots x_\ell^{\varepsilon_\ell}$, $x_i \in X$, $\varepsilon_i = \pm 1$, $\ell \geq 1$, be an element of $R \cap S$ such that ℓ is least with respect to the property that $w \notin [R \cap S, R \cap S]$ and $\beta_{12}(w) = 0$. Using the matrix multiplication we first observe that

$$\begin{aligned} 0 &= \beta_{12}(xx^{-1}) \\ &= \beta_{11}(x)\beta_{12}(x^{-1}) + \beta_{12}(x)\beta_{22}(x^{-1}) \\ &= xR\ \beta_{12}(x^{-1}) + \lambda_x\ x^{-1}S. \end{aligned}$$

This yields, for $\varepsilon = \pm 1$,

$$\beta_{12}(x^{\varepsilon}) = \varepsilon(x^{(\varepsilon-1)/2}R)\lambda_x(x^{(\varepsilon-1)/2}S).$$

Further, for $f \in F$, $x \in X$, $\varepsilon = \pm 1$, we have

$$\begin{aligned}\beta_{12}(fx^{\varepsilon}) &= \beta_{11}(f)\beta_{12}(x^{\varepsilon}) + \beta_{12}(f)\beta_{22}(x^{\varepsilon})\\ &= \varepsilon(f\ x^{(\varepsilon-1)/2}R)\lambda_x(x^{(\varepsilon-1)/2}S) + \beta_{12}(f)x^{\varepsilon}S.\end{aligned}$$

Thus by a simple induction on $\ell \geq 1$ we obtain

1.7 *FORMULA*. If $w = x_1^{\varepsilon_1}\dots x_\ell^{\varepsilon_\ell}$, $\varepsilon_i = \pm 1$, $\ell \geq 1$, then

$$\beta_{12}(w) = \sum_{j=1}^{\ell} \alpha_j(a_jR\ \lambda_{x_j}\ b_jS),$$

where,

$$a_j = x_1^{\varepsilon_1}\dots x_{j-1}^{\varepsilon_{j-1}}x_j^{(\varepsilon_j-1)/2},\quad b_j = x_j^{(\varepsilon_j-1)/2}\ x_{j+1}^{\varepsilon_{j+1}}\dots x_\ell^{\varepsilon_\ell}.$$

Now, $\beta_{12}(w) = 0$ implies that the sum of terms involving λ_x in $\beta_{12}(w)$ is zero for each $x \in \{x_1,\dots,x_\ell\}$. Thus for each $p \in \{1,\dots,\ell\}$, there exists $q \in \{1,\dots,\ell\}$, such that $x_p = x_q$, $\varepsilon_p = -\varepsilon_q$ and $a_p \equiv a_q \pmod R$, $b_p \equiv b_q \pmod S$. We may assume $p < q$. Then $a_pR = a_qR$ and $b_pS = b_qS$ yield

$$x_1^{\varepsilon_1}\dots x_{p-1}^{\varepsilon_{p-1}}x_p^{(\varepsilon_p-1)/2} \equiv x_1^{\varepsilon_1}\dots x_{q-1}^{\varepsilon_{q-1}}x_q^{(\varepsilon_q-1)/2} \pmod R$$

and

$$x_p^{(\varepsilon_p-1)/2}\ x_{p+1}^{\varepsilon_{p+1}}\dots x_\ell^{\varepsilon_\ell} \equiv x_q^{(\varepsilon_q-1)/2}\ x_{q+1}^{\varepsilon_{q+1}}\dots x_\ell^{\varepsilon_\ell} \pmod S.$$

Taking quotients and using $x_p = x_q$, $\varepsilon_p = -\varepsilon_q$ yield

$$x_{p+1}^{\varepsilon_{p+1}}\dots x_{q-1}^{\varepsilon_{q-1}} \in R \cap S.$$

It follows that for each p there exists q with $p < q$ such that the segment $x_{p+1}^{\varepsilon_{p+1}}\dots x_{q-1}^{\varepsilon_{q-1}}$ lies in $R \cap S$. Let (p,q) be a pair with the above property such that q is least possible. Then there exists a pair (p',q') such that

$$w = w_1 w_p^{\varepsilon_p} w_2 x_{p'}^{\varepsilon_{p'}} x_q^{\varepsilon_q} w_3 x_{q'}^{\varepsilon_{q'}} w_4$$

and the segments $w_2 x_{p'}^{\varepsilon_{p'}}$ and $x_q^{\varepsilon_q} w_3$ both lie in $R \cap S$. Thus modulo $[R \cap S, R \cap S]$,

$$w \equiv w_1 x_p^{\varepsilon_p} x_q^{\varepsilon_q} w_3 w_2 x_{p'}^{\varepsilon_{p'}} x_{q'}^{\varepsilon_{q'}} w_4$$

$$\equiv w_1 w_3 w_2 w_4 \ ,$$

since $x_p = x_q$, $x_{p'} = x_{q'}$, $\varepsilon_p = -\varepsilon_{q'}$, $\varepsilon_{p'} = -\varepsilon_{q'}$. Further, $\beta_{12}(w_1 w_3 w_2 w_4) = \beta_{12}(w) = 0$. Since $w_1 w_3 w_2 w_4$ is a word of length shorter than ℓ we have obtained a contradiction to our choice of w. Thus $w \in [R \cap S, R \cap S]$ as was to be proved. □

1.8 *COROLLARY*. (Enright 1968). $F/[R \cap S, R \cap S]$ is isomorphic to the group generated by all matrices $\begin{bmatrix} xR & \lambda_x \\ 0 & xS \end{bmatrix}$, $x \in X$. □

1.9 *COROLLARY*. (Schumann 1935). $F \cap (1+\mathfrak{f}\mathfrak{r}) = R' = F \cap (1+\mathfrak{r}\mathfrak{f})$. □

1.10 *COROLLARY*. (Magnus 1939). F/R' is isomorphic to the group generated by all matrices $\begin{bmatrix} xR & \lambda_x \\ 0 & 1R \end{bmatrix}$, $x \in X$. □

1.11 *REMARKS*. With $\mathfrak{x} = 0$, the proof of Theorem 1.4 shows that, in $\mathbb{Z}F$, $\sum_x u_x(x-1) = 0$ if and only if $u_x = 0$ for all x, i.e., $\mathfrak{f}$ is a free left $\mathbb{Z}F$-module with basis $\{x-1;\ x \in X\}$. Similarly, it can be shown that $\mathfrak{f}$ is a free right $\mathbb{Z}F$-module with basis $\{x-1;\ x \in X\}$. In particular, modulo $\mathfrak{f}^{n+1}$, $\mathfrak{f}^n$, $n \geq 1$, is a free abelian group with basis consisting of distinct elements $(x_1-1)\ldots(x_n-1)$, $x_i \in X$. □

1.12 *PROPOSITION*. (Gruenberg 1970, page 32). If $R \trianglelefteq F$ and $Y \subset F$ is a basis of R, then $\mathfrak{r}$ is a free left $\mathbb{Z}F$-module on $\{y-1;\ y \in Y\}$. (Similarly, $\mathfrak{r}$ is a free right $\mathbb{Z}F$-module on $\{y-1;\ y \in Y\}$.)

Proof. Let $\sum_y u_y(y-1) = 0$, $u_y \in \mathbb{Z}F$. Let T be a left transversal of R in F and write $u_y = \sum_j t_j u_{yj}$, $t_j \in T$, $u_{yj} \in \mathbb{Z}R$. Substituting in the given expression and setting the coefficient of each t_j to be zero shows that $\sum_y u_{yj}(y-1) = 0$ for each j. Since $u_{yj} \in \mathbb{Z}R$ and $\mathfrak{r}$ is a left $\mathbb{Z}R$-module on $\{y-1;\ y \in Y\}$

(Remark 1.11), it follows that $u_{yj} = 0$ in $\mathbb{Z}R$ and consequently $u_y = 0$ for all y. □

Since the map $R \to \mathfrak{r}/\mathfrak{r}\mathfrak{f}$ $(R \to \mathfrak{r}/\mathfrak{f}\mathfrak{r})$ given by $r \to (r-1) + \mathfrak{r}\mathfrak{f}$ $(r \to (r-1) + \mathfrak{f}\mathfrak{r})$ is an epimorphism, Corollary 1.9 also yields

1.13 *COROLLARY*. $R/R' \cong \mathfrak{r}/\mathfrak{r}\mathfrak{f} \cong \mathfrak{r}/\mathfrak{f}\mathfrak{r}$. □

We conclude this section by deriving the following more general result from Theorem 1.6.

1.14 *THEOREM*. (Bergman & Dicks 1975). Let $H, K \trianglelefteq G$, $\Delta(G) = \mathbb{Z}G(G-1)$, $\Delta(G,H) = \mathbb{Z}G(H-1)$ and $\Delta(G,K) = \mathbb{Z}G(K-1)$. Then $G \cap (1 + \Delta(G,H)\Delta(G,K)) = [H \cap K, H \cap K]$.

Proof. With $G = F/T$, $H = R/T$, $K = S/T$, the proof consists in showing that $F \cap (1+\mathfrak{r}\mathfrak{s}+\mathfrak{t}) = [R \cap S, R \cap S]T$, where $\mathfrak{r} = \mathbb{Z}F(R-1)$, $\mathfrak{s} = \mathbb{Z}F(S-1)$ and $\mathfrak{t} = \mathbb{Z}F(T-1)$. Indeed, let $w \in F$ be such that $w-1 \in \mathfrak{r}\mathfrak{s}+\mathfrak{t}$. Since $\mathfrak{r}\mathfrak{s}+\mathfrak{t} \in \mathfrak{r} \cap \mathfrak{s}$, it follows that $w-1 \in \mathfrak{r} \cap \mathfrak{s}$ and consequently $w \in R \cap S \le RS$. Also since $R \cap (1+\mathfrak{r}) = R \cap (1+\mathbb{Z}R(R-1))$ and $S \cap (1+\mathfrak{s}) = S \cap (1+\mathbb{Z}S(S-1))$, it follows that we may assume $F = RS$ without loss of generality.

We first prove that modulo $\mathfrak{r}\mathfrak{s}$ every element of $\mathfrak{t}$ is of the form $t-1$ for some $t \in T$. Let $u \in \mathfrak{t}$. Then u is of the form

$$
\begin{aligned}
u &= \sum_i n_i(t_i-1) + \sum_j m_j f_j(t_j-1), \quad n_i, m_j \in \mathbb{Z},\ f_j \in F,\ t_k \in T \\
&\equiv \sum_i n_i(t_i-1) + \sum_j m_j s_j(t_j-1) \pmod{\mathfrak{r}\mathfrak{s}}, \quad s_j \in S, \text{ since } F = RS \\
&\equiv \sum_i n_i(t_i-1) + \sum_j m_j(t_j^*-1)s_j \pmod{\mathfrak{r}\mathfrak{s}}, \quad t_j^* = s_j t_j s_j^{-1} \in T \\
&\equiv \sum_i n_i(t_i-1) + \sum_j m_j(t_j^*-1) \pmod{\mathfrak{r}\mathfrak{s}} \\
&\equiv t-1 \pmod{\mathfrak{r}\mathfrak{s}},
\end{aligned}
$$

where $t = \prod_i t_i^{n_i} \prod_j t_j^{*\,m_j} \in T$.

Thus $w-1 \in \mathfrak{r}\mathfrak{s}+\mathfrak{t}$ implies that $w-1 \equiv t-1 \bmod \mathfrak{r}\mathfrak{s}$ for some $t \in T$. This, in turn, yields $wt^{-1}-1 \in \mathfrak{r}\mathfrak{s}$ and consequently $wt^{-1} \in [R \cap S, R \cap S]$ by Theorem 1.6. Thus $w \in [R \cap S, R \cap S]T$ as was to be proved. □

1.15 *COROLLARY*. $G \cap (1 + \Delta(G,H)\Delta(G)) = G \cap (1 + \Delta(G)\Delta(G,H)) = H'$ for all $H \trianglelefteq G$. □

2. FREE DIFFERENTIAL CALCULUS

Let G be a group, $\mathbb{Z}G$ its integral group ring and M a left G-module. An additive map $d\colon \mathbb{Z}G \to M$ is called a <u>left derivation</u> if

$$d(xy) = x\, d(y) + d(x)$$

holds for all $x,y \in G$. A left derivation has the following properties which are easily verified.

2.1 *PROPOSITION.* (i) $d(uv) = ud(v) + \varepsilon(v)d(u)$, $u,v \in \mathbb{Z}G$ and $\varepsilon\colon \mathbb{Z}G \to \mathbb{Z}$, the unit augmentation map;
(ii) $d(1) = 0$;
(iii) $d(g^{-1}) = -g^{-1}d(g)$, $g \in G$. □

Consider the ring homomorphism

$$\psi^*\colon \mathbb{Z}F \to \begin{bmatrix} \mathbb{Z}F & \sum_x \mathbb{Z}F\,\lambda_x \\ 0 & \mathbb{Z} \end{bmatrix}$$

given by Notation 1.2 with $\underline{\underline{x}} = 0$. For $u \in \mathbb{Z}F$, set

$$\partial_x(u) = \text{coefficient of } \lambda_x \text{ in } \alpha_{12}(u). \tag{1}$$

Then $\alpha_{12}(uv) = \alpha_{11}(u)\alpha_{12}(v) + \alpha_{12}(u)\alpha_{22}(v) = u\alpha_{12}(v) + \alpha_{12}(u)\varepsilon(v)$ and it follows that $\partial_x\colon \mathbb{Z}F \to \mathbb{Z}F$ is a left derivation. This derivation has the additional property that for $x,y \in X$,

$$\partial_x(y) = \begin{cases} 1 & \text{if } x = y \\ 0 & \text{if } x \neq y. \end{cases} \tag{2}$$

If $u \in \mathbb{Z}F$ then $u-\varepsilon(u) \in \underline{\underline{\mathfrak{f}}}$ and so we can write

$$u - \varepsilon(u) = \sum_x u_x(x-1),\ u_x \in \mathbb{Z}F,\ x \in X. \tag{3}$$

If $d\colon \mathbb{Z}F \to \mathbb{Z}F$ is any left derivation then applying it to both sides of (3) shows that

$$d(u) = \sum_x u_x\, d(x). \tag{4}$$

This shows that if two derivations agree for each $x \in X$, then they agree on $\mathbb{Z}F$. Thus, for each $x \in X$, there is a unique derivation ∂_x satisfying (2). Applying ∂_x to (3) gives

$$\partial_x(u) = u_x \tag{5}$$

which together with (4) yields

$$d(u) = \sum_x \partial_x(u)\, d(x) \tag{6}$$

for $u \in \mathbb{Z}F$; and by (3) and (5),

$$u - \varepsilon(u) = \sum_x \partial_x(u)\,(x-1), \quad u \in \mathbb{Z}F .$$

The foregoing discussion brings out the fundamental properties of the left derivation and we record them as follows.

2.2 *THEOREM*. (Fox 1953). Let F be a free group with basis X. Then

(a) For each $x \in X$ there is a unique left derivation $\partial_x\colon \mathbb{Z}F \to \mathbb{Z}F$ satisfying

$$\partial_x(y) = \begin{cases} 1 & \text{if } x = y \\ 0 & \text{if } x \neq y,\ y \in X; \end{cases}$$

(b) If $d\colon \mathbb{Z}F \to \mathbb{Z}F$ is any left derivation then $d(u) = \sum_x \partial_x(u) d(x)$ for all $u \in \mathbb{Z}F$;

(c) For $u \in \mathbb{Z}F$, $u-\varepsilon(u) = \sum_x \partial_x(u)(x-1)$. □

2.3 *REMARKS*. (i) The left derivation $\partial_x\colon \mathbb{Z}F \to \mathbb{Z}F$ is called the left partial derivation with respect to x.

(ii) The formula 2.2(c) is commonly known as the Fox's fundamental formula.

(iii) Right partial derivations $\partial'_x\colon \mathbb{Z}F \to \mathbb{Z}F$ may similarly be defined by expressing $u-\varepsilon(u)$ as $\sum_x (x-1)u'(x)$ and setting $u'(x) = \partial'_x(u)$.

(iv) If $w = x_1^{\varepsilon_1} \cdots x_\ell^{\varepsilon_\ell}$, $\varepsilon_i = \pm 1$, $\ell \geq 1$, then Formula 1.7 with $R = \{1\} = S$ yields

$$\partial_{x_j}(w) = \varepsilon_j\, x_1^{\varepsilon_1} \cdots x_{j-1}^{\varepsilon_{j-1}} x_j^{(\varepsilon_j - 1)/2}$$

and

$$\partial'_{x_j}(w) = \varepsilon_j\, x_j^{(\varepsilon_j-1)/2}\, x_{j+1}^{\varepsilon_{j+1}} \cdots x_\ell^{\varepsilon_\ell}. \qquad \square$$

Let $R \trianglelefteq F$ and let $\theta\colon \mathbb{Z}F \to \mathbb{Z}(F/R) \cong \mathbb{Z}F/\underline{r}$ be the natural projection. We prove,

2.4 *THEOREM*. (Fox 1953). $R' = \{f \in F \mid \theta\partial_x(f) = 0$ for all $x \in X\}$.

Proof. If $\theta\partial_x(f) = 0$ for all $x \in X$ then $\partial_x(f) \in \underline{r}$ for all $x \in X$. By Theorem 2.2(c), $f-1 = \sum_x \partial_x(f)(x-1) \in \underline{\underline{rf}}$ and, in turn, $f \in R'$ by Corollary 1.9.

Conversely, let $f \in R'$. Then $f-1 \in \underline{r}^2 \leq \underline{\underline{rf}}$. Since $f-1 = \sum_x \partial_x(f)(x-1)$, it follows by Remark 1.11 that $\partial_x(f) \in \underline{r}$ for all $x \in X$. This yields $\theta\partial_x(f) = 0$ for all $x \in X$. □

The Magnus representation

$$\psi^*: \mathbb{Z}F \to \begin{bmatrix} \mathbb{Z}F/\underline{r} & \sum_x (\mathbb{Z}F/\underline{r})\lambda_x \\ 0 & \mathbb{Z}+\underline{r}/\underline{r} \end{bmatrix}$$

yields the homomorphism

$$\alpha_{12}: R/R' \to \sum_x \mathbb{Z}(F/R)\lambda_x$$

and it is desirable to know a criterion to determine which elements of $\sum_x \mathbb{Z}(F/R)\lambda_x$ actually occur as images of elements of R/R'. This is achieved by the following occurrence theorem.

2.5 *THEOREM*. (Remeslennikov & Sokolov 1970). Let $u_{x(1)},\dots,u_{x(n)} \in \mathbb{Z}F$. Then $\alpha_{12}(u) = \sum_i \theta\, u_{x(i)}\lambda_{x(i)}$ for some $u \in R$ if and only if we have $\theta \sum_i u_{x(i)}(x(i)-1) = 0$ in $\mathbb{Z}(F/R)$.

Proof. If $\theta \sum_i u_{x(i)}(x(i)-1) = 0$ in $\mathbb{Z}(F/R)$ then $\sum_i u_{x(i)}(x(i)-1) = v \in \underline{r}$ and we may write $v \equiv \sum_i n_i(r_i-1) \pmod{\underline{\underline{rf}}}$. Put $u = \prod_i r_i^{n_i} \in R$. Then $\alpha_{12}(u) = \alpha_{12}(v) = \sum_i \theta\, u_{x(i)}\lambda_{x(i)}$ with $\partial_{x(i)}(u) = \partial_{x(i)}(v) = u_{x(i)}$ for all i.

Conversely, let $u \in R$ with $\alpha_{12}(u) = \sum_{i=1}^{n} \theta\, u_{x(i)}\lambda_{x(i)}$. Then $\theta\partial_{x(i)}(u) = \theta u_{x(i)}$ and $\theta\partial_x(u) = 0$ for $x \notin \{x(1),\dots,x(n)\}$. Since $u = \sum_x \partial_x(u)(x-1)$, we have

$$0 = \theta(u-1) = \sum_x \theta(\partial_x(u)(x-1))$$

$$= \sum_i \theta(\partial_{x(i)}(u)(x(i)-1))$$

$$= \theta \sum_i u_{x(i)}(x(i)-1). \qquad \square$$

In Theorem 2.4 we have seen that for $f \in F$, $\theta\partial_x(f) = 0$ for all $x \in X$ if and only if $f \in R'$. This led Fox to raise the following problem.

2.6 *FOX PROBLEM*. Given $n \geq 1$ and the natural projection $\theta\colon F \to F/R$, identify the elements $f \in F$ which satisfy $\theta\partial_{x(1)}\cdots\partial_{x(k)}(f) = 0$ for all $x(i) \in X$ and all $1 \leq k \leq n$. $\square$

An iteration of the Fox's fundamental formula (Theorem 2.2(c)) shows that for $u \in \mathbb{Z}F$ and $n \geq 1$,

$$\begin{aligned} u = \varepsilon(u) &+ \sum_x (\varepsilon\, \partial_x(u))(x-1) \\ &+ \sum_{x(1),x(2)} (\varepsilon\, \partial_{x(1)}\partial_{x(2)}(u))(x(1)-1)(x(2)-1) \\ &+ \cdots \\ &+ \sum_{x(1),\ldots,x(n-1)} (\varepsilon\, \partial_{x(1)}\cdots\partial_{x(n-1)}(u)) (x(1)-1)\cdots(x(n-1)-1) \\ &+ \sum_{x(1),\ldots,x(n)} \partial_{x(1)}\cdots\partial_{x(n)}(u)(x(1)-1)\cdots (x(n)-1). \end{aligned} \tag{7}$$

In particular, if $f \in F$ and $\theta\partial_{x(1)}\cdots\partial_{x(k)}(f) = 0$ for all $x(i) \in X$ and all $1 \leq k \leq n$, then $\varepsilon\partial_{x(1)}\cdots\partial_{x(k)}(f)$ $= \varepsilon\ \theta\ \partial_{x(1)}\cdots\partial_{x(k)}(f) = 0$ for all $1 \leq k \leq n-1$ and $\partial_{x(1)}\cdots\partial_{x(n)}(f) \in \underline{\underline{\mathfrak{r}}}$. Thus $f-1 \in \underline{\underline{\mathfrak{r}\mathfrak{f}}}^n$ and conversely. Hence the Fox problem is equivalent to the following problem.

2.7 *FOX PROBLEM*. (An alternate form). Identify the normal subgroups $F(n,R) = F \cap (1+\underline{\underline{\mathfrak{r}\mathfrak{f}}}^n)$ for $n \geq 1$. $\square$

We call $F(n,R)$ the <u>n-th Fox subgroup of F relative to R</u>. In Corollary 1.9 we have identified $F(1,R)$ as R'. The identification for $n \geq 2$ is discussed in Chapter III.

For a group G let $\gamma_n(G)$ be the n-th term of its lower central series defined inductively by: $\gamma_1(G) = G$ and, for $n \geq 2$, $\gamma_n(G) = [\gamma_{n-1}(G), G] = \langle [g,h];\ g \in \gamma_{n-1}(G),\ h \in G \rangle$, where $[g,h] = g^{-1}h^{-1}gh$. It is readily seen that $\gamma_n(G)$ is generated by all left-normed commutators $[g_1,\ldots,g_n]$ $(= [[\ldots[g_1,g_2],\ldots],g_n])$, $g_i \in G$ (see, for instance, Magnus et al. 1966). For $g,h \in G$ we have, in $\mathbb{Z}G$,

$$([g,h]-1) = g^{-1}h^{-1}((g-1)(h-1) - (h-1)(g-1))$$

and it follows that $\gamma_2(G) \leq G \cap (1+\Delta^2(G))$, where $\Delta(G) = \mathbb{Z}G(G-1)$ is the augmentation ideal of $\mathbb{Z}G$. More generally, for all n, we have by a simple induction, $\gamma_n(G) \leq G \cap (1+\Delta^n(G)) = D_n(G)$, the n-th integral dimension subgroup of G. The study of the structure of the quotients $D_n(G)/\gamma_n(G)$ is known as the <u>dimension subgroup problem</u>. In the language of free group rings the dimension subgroup problem, with $G = F/R$, is equivalent to the following problem.

2.8 *DIMENSION SUBGROUP PROBLEM*. (An alternate form). Identify the normal subgroups $D(n,R) = F \cap (1+\underline{\underline{r}}+\underline{\underline{f}}^n)$ for $n \geq 1$. □

We call $D(n,R)$ the <u>n-th dimension subgroup of F relative to R</u>. We shall study the structure of $D(n,R)$ in Chapter IV. Here, in the next section, we prove that $F \cap (1+\underline{\underline{f}}^n) = \gamma_n(F)$ for all $n \geq 1$. This identification may be regarded as the first major result in free group rings. [Note that $F(n-1,F) = F \cap (1+\underline{\underline{f}}^n) = D(n,\{1\})$.]

3. THE FUNDAMENTAL THEOREM OF FREE GROUP RINGS

We recall Magnus embedding of a free group into a ring of formal power series. Let $F = \langle x_1, x_2, \ldots \rangle$ be a free group of finite or countably infinite rank and let $\mathcal{P} = \mathbb{Z}[[a_1,a_2,\ldots]]$ be the ring of formal power series over $\mathbb{Z}$ in the non-commuting indeterminates $a_1,a_2,\ldots$. The elements $1+a_i$ are invertible with

$$(1+a_i)^{-1} = 1 - a_i + a_i^2 - a_i^3 + \ldots .$$

The map $\beta\colon x_i \to 1+a_i$, $1 \leq i$, extends to a homomorphism of F into $U(\mathcal{P})$, the group of units of $\mathcal{P}$. Let $w = x_{i(1)}^{\alpha(1)} \ldots x_{i(\ell)}^{\alpha(\ell)}$, $\ell \geq 1$, $\alpha(i) \in \mathbb{Z}\setminus\{0\}$, $x_{i(j)} \neq x_{i(j+1)}$, be a freely reduced word in F. Then in the power series $\beta(w)$ the coefficient of the monomial $a_{i(1)}\cdots a_{i(\ell)}$ is easily seen to be $\alpha(1)\ldots\alpha(\ell) \neq 0$. It follows that $\beta\colon F \to U(P)$ is a monomorphism. We extend β by linearity to a ring homomorphism

$$\beta^*\colon \mathbb{Z}F \to P$$

and denote by J the ideal of P generated by $a_1, a_2, \ldots$. Following Magnus (1937) we define

$$\mathcal{D}_n(F) = \{w \in F \mid \beta^*(w) \in 1+J^n,\ n \geq 1\}. \tag{8}$$

Then we have,

3.1 *PROPOSITION*. $\mathcal{D}_n(F) = F \cap (1+\underline{\underline{f}}^n)$.

Proof. Clearly $\beta^*(\underline{\underline{f}}^n) \subseteq J^n$ and it follows that $F \cap (1+\underline{\underline{f}}^n) \leq \mathcal{D}_n(F)$. Conversely, let $w \in \mathcal{D}_n(F)$ so that $\beta^*(w-1) \in J^n$. With $u = w-1$, applying β^* to both sides of (7) shows that, for $1 \leq k \leq n-1$, the coefficient of $a(1)\ldots a(k)$ in the power series $\beta^*(u)$ is $\varepsilon\partial_{x(1)}\cdots\partial_{x(k)}(u)$ which is zero by hypothesis. It follows that $u \in \underline{\underline{f}}^n$ and consequently $w \in F \cap (1+\underline{\underline{f}}^n)$. □

[Proposition 3.1 is included to point out a connection between free group rings and the ring of formal power series. We shall not need it for our proof of the fundamental theorem.]

Let $\underline{\underline{A}} = \mathbb{Z}[A]$ be the free associative algebra over $\mathbb{Z}$ generated by a set $A = \{a_1, a_2, \ldots\}$ and let X be an algebraically independent subset of $\underline{\underline{A}}$ (i.e., $\mathbb{Z}[X]$ is a free subalgebra of $\underline{\underline{A}}$). Regarding $\underline{\underline{A}}$ as a Lie ring under $((u,v)) = uv-vu$, let $\underline{\underline{L}}(X)$ denote the Lie subring of $\underline{\underline{A}}$ generated by X. The elements of $\underline{\underline{L}}(X)$ will be referred to as the Lie elements (in X). The Lie ring $\underline{\underline{L}}(X)$ satisfies the important Jacobi identity:

$$((u,v,w)) + ((v,w,u)) + ((w,u,v)) = 0. \tag{9}$$

Using (9) it follows that every element of $\underline{\underline{L}}(X)$ is a $\mathbb{Z}$-linear sum of left-normed Lie elements of weight n of the form $((x_1,\ldots,x_n))$, $n \geq 1$, $x_i \in X$. Furthermore, using (9) it can be readily verified that a left-normed Lie element of weight n with an entry x can be expressed as a $\mathbb{Z}$-linear sum of left-normed Lie elements of weight n each of which has the first entry x. These elementary facts will be used without further reference. For each $m \geq 1$, let $L_m(X)$ denote the (additive) subgroup of $\underline{L}(X)$ generated by all Lie elements (in X) of weight m. Then

$$\underline{L}(X) = \sum_{m=1}^{\infty} \oplus L_m(X). \tag{10}$$

If $u = ((x_1,\ldots,x_n)) \in L_n(X)$, then u is a $\mathbb{Z}$-linear sum of monomials $x_{1\sigma}\ldots x_{n\sigma}$ of <u>length</u> n where σ is a permutation of $\{1,\ldots,n\}$. The underlying set $\{x_1,\ldots,x_n\}$ is referred to as <u>the support of u in X</u> and we write

$$\text{Support}_X(u) = \{x_1,\ldots,x_n\}.$$

A straightforward monomial expansion of $u = ((x, y_1,\ldots,y_q))$, $q \geq 2$, $x, y_i \in X$, yields

$$u = xy_1\ldots y_q + (-1)^q y_q\ldots y_1 x + u' , \tag{11}$$

where u' is a linear sum of monomials of the form $y_{1\sigma}\ldots y_{i\sigma} x y_{(i+1)\sigma}\ldots y_{q\sigma}$ with $1 \leq i \leq q-1$.

Similarly, expansion of $v = ((x, y_1,\ldots,y_q,\underbrace{x,\ldots,x}_{p}))$ $= ((u, \underbrace{x,\ldots,x}_{p}))$, $p \geq 1$, yields

$$v = (-1)^p x^{p+1} y_1\ldots y_q + (-1)^q y_q\ldots y_1 x^{p+1} + v' \tag{12}$$

where v' is a linear combination of summands $x^i u x^j$, $i+j = p$, $i,j \geq 1$ for any $p \geq 2$.

Suppose now that X is an algebraically independent and totally-ordered subset of $\underline{A}$ $(= \mathbb{Z}[A])$. For each $x \in X$, we define a set $B_X(x)$ of certain Lie elements as follows:

$$B_X(x) = \{ ((x, y_1,\ldots,y_q,\underbrace{x,\ldots,x}_{p})) \mid p \geq 0,\ q \geq 1,\ x < y_i \in X,\ i = 1,\ldots,q\}. \tag{13}$$

For $u = ((x, y_1,\ldots,y_q,\underbrace{x,\ldots,x}_{p}))$, $v = ((x, z_1,\ldots,z_{q'},\underbrace{x,\ldots,x}_{p'}))$, we define $u < v$ if either $p < p'$ or $p = p'$ and $(y_1,\ldots,y_q) < (z_1,\ldots,z_{q'})$ lexicographically. This defines a total order on $B_X(x)$.

3.2 <u>*LEMMA*</u>. $B_X(x)$ is an algebraically independent subset of $\underline{A}$

<u>*Proof*</u>. Let $u_1 < \ldots < u_n$, $n \geq 1$, be non-zero elements of $B_X(x)$

with $u_1\rho_1 + \dots + u_n\rho_n = 0$ for some $\rho_i \in \mathbb{Z}[B_X(x)]\backslash\{0\}$. To complete the proof it suffices to show that each ρ_i is zero. The proof is by induction on n. If $n = 1$ then $u_1\rho_1 = 0$ is contrary to the fact that $\underset{=}{A}$ has no zero divisors. Thus we may assume that $n \geq 2$ and further that each u_i has the form

$$u_i = ((x, y_{i1},\dots,y_{iq(i)},\underbrace{x,\dots,x}_{\leftarrow p_i \rightarrow})) \tag{14}$$

with $0 \leq p_1 \leq \dots \leq p_n$.

If, for some $1 \leq k < n$, $p_k < p_{k+1} = \dots = p_n = p$, say, then the monomial expansion of $u_{k+1}\rho_{k+1} + \dots + u_n\rho_n$ yields, by (12), a summand

$$\alpha = x^{p+1}y_{(k+1)1}\cdots y_{(k+1)q(k+1)}\rho_{k+1} + \dots$$
$$\dots + x^{p+1}y_{n1}\cdots y_{nq(n)}\rho_n$$

whose terms do not cancel in the expansion of $u_1\rho_1 + \dots + u_k\rho_k$. Thus we must have $\alpha = 0$ which, in turn, implies that

$$y_{(k+1)1}\cdots y_{(k+1)q(k+1)}\rho_{k+1} + \dots$$
$$\dots + y_{n1}\cdots y_{nq(n)}\rho_n = 0. \tag{15}$$

Since $u_{k+1} < \dots < u_n$, it follows that (15) has the form

$$\mu_1\rho_{k+1} + \mu_1\mu_2\rho_{k+2} + \dots + \mu_1\cdots\mu_{n-k}\rho_n = 0$$

where $\mu_i \in \mathbb{Z}[Y]\backslash\{0\}$, $x \notin Y$. Thus

$$\rho_{k+1} + \mu_2\rho_{k+2} + \dots + \mu_2\cdots\mu_{n-k}\rho_n = 0. \tag{16}$$

Since $\rho_{k+1} \neq 0$, it has, in its monomial expansion, non-zero terms beginning with x which obviously cannot cancel in $\mu_2\rho_{k+2} + \dots + \mu_2\cdots\mu_{n-k}\rho_n$. Thus (16) is not valid and we must have $p_1 = p_2 = \dots = p_n = p$. However, as before, we obtain

$$x^{p+1}\mu_1\rho_1 + x^{p+1}\mu_1\mu_2\rho_2 + \dots + x^{p+1}\mu_1\cdots\mu_n\rho_n = 0$$

which yields

$$\rho_1 + \mu_2\rho_2 + \dots + \mu_2\cdots\mu_n\rho_n = 0$$

and, in turn, $\rho_1 = 0$ contrary to our choice. This completes the proof. □

3.3 *LEMMA*. Let $x_1 < \dots < x_n \in X$ and $u_i \in \mathbb{Z}[B_X(x_i)]$ for $i = 1,\dots,n$. Then $u_1,\dots,u_n$ are $\mathbb{Z}$-linearly independent.

Proof. Let $t_1u_1 + \dots + t_nu_n = 0$ for some $t_i \in \mathbb{Z}$. Since $x_1 \in \text{Support}_X(u_1)$ and $x_1 \notin \text{Support}_X(u_i)$, $i \geq 2$, it follows by the algebraic independence of X that $t_1 = 0$ and, likewise, $t_2 = \dots = t_n = 0$. □

The Lemmas 3.2 and 3.3 provide a tool for constructing a linearly independent totally-ordered set $B(A)$ of Lie elements (in A) which generate $\underline{L}(A)$. We proceed as follows:

Set $B(\phi) = A = \{a_1,a_2,\dots\} = \{b_{i(1)};\ i(1) \in I(\phi)\}$, where $I(\phi)$ is an ordered indexing set for $B(\phi)$. For each $i(1) \in I(\phi)$ we use (13) and Lemma 3.2 to construct the algebraically independent and totally ordered set $B(i(1)) = B_X(x)$ with $x = b_{i(1)}$ and $X = B(\phi)$.

Let $B(i(1)) = \{b_{i(2)};\ i(2) \in I(i(1))\}$, where $I(i(1))$ is an ordered indexing set for $B(i(1))$. For each $i(2) \in I(i(1))$ we construct as before the algebraically independent and totally-ordered set $B(i(1),i(2)) = B_X(x)$ with $x = b_{i(2)}$ and $X = B(i(1))$.

Having constructed and totally ordered $B(i(1),\dots,i(k)) = \{b_{i(k+1)};\ i(k+1) \in I(i(1),\dots,i(k))\}$, where $I(i(1),\dots,i(k))$ is an ordered indexing set for $B(i(1),\dots,i(k))$, $k \geq 2$, we construct for each $i(k+1) \in I(i(1),\dots,i(k))$ the algebraically independent and totally-ordered set $B(i(1),\dots,i(k+1)) = B_X(x)$ with $x = b_{i(k+1)}$, $X = B(i(1),\dots,i(k))$.

Put

$$B(A) = \bigcup_{k\geq 0} B(i(1),\dots,i(k)) \tag{17}$$

where the union is formed over all k-tuples $(i(1),\dots,i(k))$, $i(j)$ running over the indexing set for $B(i(1),\dots,i(j-1))$.

For $u,v \in B(A)$ we say that $u < v$ provided $u \in B(i(1),\dots,i(k))$, $v \in B(j(1),\dots,j(\ell))$, $k,\ell \geq 1$, and either, under the lexicographic ordering, $(i(1),\dots,i(k)) < (j(1),\dots,j(\ell))$ or u,v both lie in $B(i(1),\dots,i(k))$ and $u < v$ as elements of $B(i(1),\dots,i(k))$. Further, if $u = b_{i(1)} \in B(\phi)$ and $v \in B(j(1),\dots,j(\ell))$ then $u < v$ if

$i(1) \leq j(1)$ and $v < u$ if $j(1) < i(1)$. With the above ordering, $B(A)$ is a totally-ordered set.

3.4 *LEMMA*. $B(A)$ is a linearly independent totally-ordered subset of $\underline{A}$.

Proof. Let $u_1 < u_2 < \ldots < u_n$, $n \geq 2$, be elements of $B(A)$ with $t_1u_1 + \ldots + t_nu_n = 0$ for some $t_i \in \mathbb{Z}\setminus\{0\}$. We may assume without loss of generality that each $u_i \in L_m(A)$ for some fixed integer m. For $X = B(i(1),\ldots,i(k))$ and $X_j = B(i(1),\ldots,i(j))$ we have $X \subset \mathbb{Z}[X_{k-1}] \subset \ldots \subset \mathbb{Z}[X_1] = \mathbb{Z}[B(i(1))]$. Thus by Lemma 3.3 we must have identical first coordinates for the representation of each u_i as an element of $B(i(1),\ldots,i(k))$. Iterating the argument we see that we need only consider the case when

$$u_1 \in B(i(1),\ldots,i(k)) = X$$

$$u_2 \in B(i(1),\ldots,i(k),i(k+1))$$

$$\vdots$$

$$u_n \in B(i(1),\ldots,i(k),i(k+1),\ldots,i(k+n-1)).$$

Then we have $t_1u_1 + \ldots + t_nu_n = 0$ and $u_1,\ldots,u_n \in \mathbb{Z}[X]$ which is a free subalgebra of $\underline{A}$. As elements of $\mathbb{Z}[X]$, $u_2,\ldots,u_n$ are elements of degree at least 2 whereas u_1 is of degree 1. Thus by the algebraic independence of X (Lemma 3.2) we conclude that $t_1u_1 = 0$, i.e., $t_1 = 0$, a contradiction. □

3.5 *LEMMA*. $B(A)$ spans $\underline{L}(A)$.

Proof. It suffices to prove by induction on $n \geq 1$ that if X is an algebraically independent totally-ordered subset of $\underline{A}$ then $B(X) \cap L_n(X)$ spans $L_n(X)$. For $n = 1$ there is nothing to prove. Let $n \geq 2$ and assume $B(X) \cap L_m(X)$ spans $L_m(X)$ for $m < n$. Let $u = ((x_{i(1)},\ldots,x_{i(n)})) \in L_n(X)$. Let $Y = \text{Support}_X(u)$ and let y be its least element. Then by repeated application of the Jacobi identity (9) we may express u as an element of $\underline{L}(Y)$ and as such $u \in L_1(Y) \oplus \ldots \oplus L_{n-1}(Y)$. By the induction hypothesis for each $m \le n-1$, $B(Y) \cap L_m(Y)$ spans $L_m(Y)$. Since $B(Y) \subseteq B(X)$ it follows that u is a linear sum of elements of $B(X) \cap L_n(X)$. □

3.6 *COROLLARY*. $\underline{\underline{L}}(A) = \sum_{n=1}^{\infty} \oplus L_n(A)$ is a free Lie algebra and $L_n(A) \cap B(A)$ is a basis of the additive group $L_n(A)$ for each $n \ge 1$.

Proof. Let Lie(A) be the free Lie ring freely generated by the set A with respect to the multiplication written as $((\ ,\ ))$. Construct $B(A)$ as is done in (17). The argument in the proof of Lemma 3.5 shows that Lie(A) is a span of $B(A)$. On the other hand, by Lemma 3.4 the image of $B(A)$ under the homomorphism η: Lie(A) $\to \underline{\underline{L}}(A)$, extending the trivial map $A \to A$, is linearly independent. Thus η is an isomorphism, proving the corollary. □

Let $F = \langle x_1,\ldots,x_m \rangle$, $m \ge 2$, be a free group. Then $\sum_{n=1}^{\infty} \oplus \gamma_n(F)/\gamma_{n+1}(F)$ is a Lie algebra generated by $x_i\gamma_2(F)$, $i = 1,\ldots,m$, where for $u \in \gamma_i(F)\setminus\gamma_{i+1}(F)$, $v \in \gamma_j(F)\setminus\gamma_{j+1}(F)$, $((u\gamma_{i+1}(F),\ v\gamma_{j+1}(F))) = [u,v]\gamma_{i+j+1}(F)$ (Lazard 1954). Let $\underline{\underline{A}} = \mathbb{Z}[A]$, $A = \{a_1,\ldots,a_m\}$, be a free associative algebra. Then $\underline{\underline{A}} = \sum_{n=0}^{\infty} \oplus A_n$, where A_n is the n-th homogeneous component of $\underline{\underline{A}}$. By Corollary 3.6 and Remark 1.11 we have the following sequence of maps (cf. Serre 1965, Chapter 4, §6)

$$\sum_{n=1}^{\infty} \oplus L_n(A) \xrightarrow{\delta_1} \sum_{n=1}^{\infty} \oplus \gamma_n(F)/\gamma_{n+1}(F)$$

$$\xrightarrow{\delta_2} \sum_{n=1}^{\infty} \oplus D_n(F)/D_{n+1}(F) \qquad (D_i(F) = D(i,\{1\}))$$

$$\xrightarrow{\delta_3} \sum_{n=1}^{\infty} \oplus \underline{\underline{\mathfrak{f}}}^n/\underline{\underline{\mathfrak{f}}}^{n+1} \tag{18}$$

$$\xrightarrow{\delta_4} \sum_{n=1}^{\infty} \oplus A_n ,$$

where $\delta_1: a_i \to x_i\gamma_2(F)$; $\delta_2: w\gamma_{n+1}(F) \to wD_{n+1}(F)$; $\delta_3: wD_{n+1}(F) \to w-1+\mathfrak{f}^{n+1}$ and $\delta_4: (x_{i_1}-1)\ldots(x_{i_n}-1)+\mathfrak{f}^{n+1} \to a_{i_1}\ldots a_{i_n}$. Moreover, the composite map $\delta_1\delta_2\delta_3\delta_4$ is a monomorphism $\eta: \mathfrak{L}(A) \to \mathfrak{A}$ induced by $a_i \to a_i$. It follows, in particular, that if $w \in \gamma_n(F)\backslash\gamma_{n+1}(F)$ then $\delta_3(\delta_2 w\gamma_{n+1}(F))$ is a non-zero element of $\mathfrak{f}^n/\mathfrak{f}^{n+1}$. Hence $w-1 \notin \mathfrak{f}^{n+1}$ and, in turn, $w \in D_n(F)\backslash D_{n+1}(F)$. We have thus proved the following theorem attributed to Magnus (1937). □

3.7 *THEOREM*. (Fundamental Theorem of Free Group Rings). $F \cap (1+\mathfrak{f}^n) = \gamma_n(F)$ for all $n \geq 1$. □

Another consequence of (18) is that δ_1 is an isomorphism. Thus together with Lemmas 3.4 and 3.5 we have proved the following important theorem for free groups.

3.8 *THEOREM*. (A Basis Theorem). For $n \geq 1$, let $B_n(F)$ denote the set of group commutators obtained from the set $B(A) \cap L_n(A)$ upon replacing each a_i by x_i and the Lie operation ((,)) by the group commutator [,]. Then module $\gamma_{n+1}(F)$, $B_n(F)$ freely generates $\gamma_n(F)$. □

3.9 *REMARKS*. Our proof of Theorem 3.7 is independent, self-contained and significantly simpler than the known proofs (cf. Baumslag (1971), Passi (1979), Huppert & Blackburn (1982), Vol. 2). For some historical comments about Theorem 3.7 we refer to Chandler & Magnus (1982) where this theorem is also attributed to Grün (1936) and Witt (1937). [In an interesting article, Röhl (1985) gives a detailed account of Grün's paper and, among other things, points out a fundamental error in Grün's proof.] The resulting Theorem 3.8 seems to be of lesser group theoretic significance than the well-known Hall Basis Theorem. For instance, our basic commutators do not preserve the various polynilpotent-types which are a vital part of the Hall basic commutators. □

Some important discussions with Frank Levin and Inder Bir Passi are gratefully acknowledged.

4. THE POINCARÉ-BIRKHOFF-WITT THEOREM

In this section we shall use the results of Section 3 and derive the important Poincaré-Birkhoff-Witt Theorem for the universal enveloping rings of arbitrary Lie rings. We begin our study with the construction of a free additive basis for $\underline{\underline{A}} = \mathbb{Z}[A]$ in terms of the totally-ordered basis $B(A)$ of the free Lie ring $\underline{\underline{L}}(A)$ as given by Lemmas 3.4 and 3.5.

Let X be an algebraically independent and totally-ordered subset of the free associative algebra $\underline{\underline{A}} = \mathbb{Z}[A]$. For $x \in X$ let $B_X(x)$ be as defined by (13). If $u = ((x,y_1,\ldots,y_q,\underbrace{x,\ldots,x}_{\leftarrow p \rightarrow}))$ $\in B_X(x)$ then we say that u is <u>distinguished</u> (in $\mathbb{Z}[X]$) by the monomial $u^* = x^{p+1}y_1\ldots y_q$. Set $B_X^+(x) = B_X(x) \cup \{x\}$ and $x^* = x$. We say that the set

$$P(x) = \{u_n\ldots u_1;\ u_1 \le \ldots \le u_n \in B_X^+(x)\}$$

of all ordered products of elements of $B_X^+(x)$ is distinguished by the set of all monomials

$$P^*(x) = \{u_n^*\ldots u_1^*;\ u_i^* \text{ distinguishing } u_i\}.$$

Using the $\mathbb{Z}$-linear independence of $P^*(x)$ it follows (as in the proof of Lemma 3.2) that $P(x)$ is a $\mathbb{Z}$-linearly independent subset of $\mathbb{Z}[X]$. Now let $x_1 < \ldots < x_k$ and define

$$P(x_1,\ldots,x_k) = \{P_k\ldots P_1;\ P_i \in P(x_i)\}.$$

Since $x_i \notin \mathrm{Support}_X\left(\bigcup_{j=i+1}^{k} B_X(x_j)\right)$, $i = 1,\ldots,k-1$, it follows that $P(x_1,\ldots,x_k)$ is linearly independent being distinguished by the set

$$P^*(x_1,\ldots,x_k) = \{P_k^*\ldots P_1^*;\ P_i^* \text{ distinguishing } P_i\}.$$

An iteration of the above procedure yields

4.1 <u>*LEMMA*</u>. Let $B(A)$ be the totally-ordered set given by (17). Then the $B(A)$-monomials of the form $b_n\ldots b_1$ with $b_1 \le \ldots \le b_n$, $b_i \in B(A)$, are $\mathbb{Z}$-linearly independent. □

We next prove,

4.2 <u>*LEMMA*</u>. The elements of the form $b_n\ldots b_1$ with

$b_1 \le \ldots \le b_n$, $b_i \in B(A)$, form an additive basis for $\underline{\underline{A}} = \mathbb{Z}[A]$.

Proof. In view of Lemma 4.1 we need only prove that every element of $\underline{\underline{A}}$ is a $\mathbb{Z}$-linear sum of ordered $B(A)$-monomials $b_n \ldots b_1$ with $b_1 \le \ldots \le b_n \in B(A)$. Since $A = B(\phi) \subset B(A)$, A is clearly spanned by the unordered $B(A)$-monomials $b_n \ldots b_1$. It suffices, therefore, to prove that every $B(A)$-monomial of length n can be written as a sum of ordered $B(A)$-monomials of length up to n. We proceed by induction on n. If $n = 1$ there is nothing to prove. Let $n \ge 2$ or assume the assertion for $B(A)$-monomials of length up to $n-1$. Let $v = b_n \ldots b_1$ be a $B(A)$-monomials of length n. If $b_1 \le \ldots \le b_n$ then v is of the required type. Otherwise, $b_i > b_{i+1}$ for some $i = 1,\ldots,n-1$. Now $b_{i+1}b_i = b_i b_{i+1} + c$, where $c = ((b_{i+1}, b_i))$ is a Lie element. By Lemma 3.5 we can write c as a linear sum of elements of $B(A)$. Thus

$$b_n \ldots b_{i+1} b_i \ldots b_1 = b_n \ldots b_i b_{i+1} \ldots b_1 + B(A)\text{-monomials of length } n-1.$$

The proof now follows by the induction hypothesis. □

We can now prove the free Lie ring version of the Poincaré-Birkhoff-Witt Theorem.

4.3 *THEOREM*. (Witt 1937). Let $(E;\le')$ be a totally-ordered free additive basis for $\underline{\underline{L}}(A)$. Then the set of all ordered E-monomials of the form $e_1 \ldots e_n$ with $e_1 \le' \ldots \le' e_n \in E$ form a free additive basis for $\underline{\underline{A}} = \mathbb{Z}[A]$.

Proof. We introduce a new total ordering "$\le^*$" of E by defining $e_1 \le^* e_2$ if and only if $e_2 \le' e_1$. Then the proof consists in proving that the set of all $*$-ordered E-monomials of the form $e_n \ldots e_1$ with $e_1 \le^* \ldots \le^* e_n \in E$ form a free additive basis for $\underline{\underline{A}}$. Since $A \subseteq \underline{\underline{L}}(A)$, $\underline{\underline{A}}$ is clearly spanned by the set of all E-monomials which can be $*$-ordered as in Lemma 4.2. Thus we need only prove the linear independence of the $*$-ordered E-monomials $e_n \ldots e_1$ with $e_1 \le^* \ldots \le^* e_n$. Let E_1 be a finite subset of E. Then it clearly suffices to prove the above assertion for E_1-monomials. Since $B = B(A)$ is a basis for $\underline{\underline{L}}(A)$ (Lemma 3.5), there exists a finite subset B_1 of B such that $\langle E_1 \rangle \le \langle B_1 \rangle$, where $\langle E_1 \rangle, \langle B_1 \rangle$ are additive groups generated by E_1, B_1 respectively. Further enlarging E_1 to a finite subset E_2 of E we find $\langle E_1\rangle \le \langle B_1\rangle \le \langle E_2\rangle = \langle E_1\rangle \oplus C$. It follows that $\langle B_1\rangle = \langle E_1\rangle \oplus C \cap \langle B_1\rangle$. Let $E_3 \subseteq E$ be a free basis for $C \cap \langle B_1\rangle$ and put $E_4 = E_1 \cup E_3$. Then $\langle B_1\rangle = \langle E_4\rangle$ and $|B_1| = |E_4|$. By Lemma 4.1 the ordered B_1-monomials of the form $b_n \ldots b_1$ with $b_1 \le \ldots \le b_n$ are linearly independent and consequently the *-ordered E_4-monomials of the form $e_n \ldots e_1$ with $e_1 \le^* \ldots \le^* e_n$ are also linearly independent. Since E_1 is a subset of E_4, we have established the required linear independence of the *-ordered E_1-monomials. This completes the proof of the theorem. □

Let L be a Lie ring. An associative ring $u(L)$ is called a <u>universal enveloping ring</u> for L if there exists a Lie ring homomorphism $\varepsilon: L \to u(L)$ satisfying the following property: If $\eta: L \to V$ is a Lie ring homomorphism of L into an associative ring V then there exists a unique homomorphism $\theta: u(L) \to V$ which makes the diagram

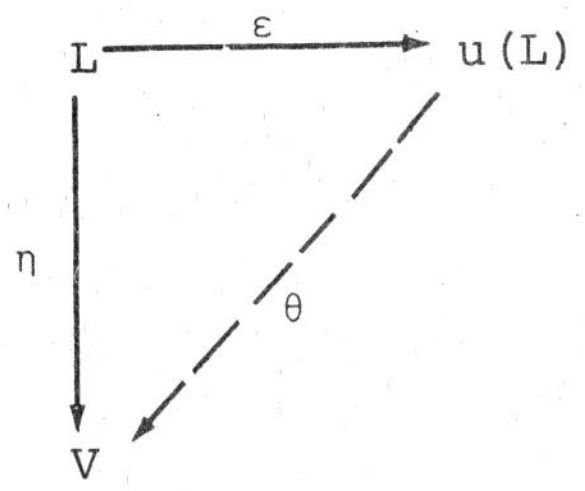

commutative (here $u(L)$ and V are regarded as Lie rings with respect to the operation $((a,b)) = ab-ba$). Clearly $u(L)$, if exists, is unique. If $L = \underline{L}(A)$, the free Lie ring associated with the free $\mathbb{Z}$-algebra $\underline{A} = \mathbb{Z}[A]$, then it is clear that $u(L) = \underline{A}$. We can now prove (cf. Jacobson 1962, Serre 1965),

4.4 *THEOREM*. (Poincaré-Birkhoff-Witt). Let E be a totally-ordered free additive basis of a Lie ring L. Then the set of all ordered E-monomials of the form $e_1 \ldots e_n$ with $e_1 \le \ldots \le e_n \in E$ forms a free additive basis for $u(L)$.

Proof. For a suitable choice of A we represent L as

$$L = \underline{L}(A)/J,\ J \ne 0$$

(if $J = 0$ the proof follows from Theorem 4.3). Consider a basis $\bar{F}$ of J and extend to a basis of $\underline{L}(A)$ by adjoining a set $\bar{E}$ of elements satisfying $\bar{e}+J = e$, $e \in E$. We introduce a total-ordering on $\bar{E} \cup \bar{F}$ so that the elements of $\bar{E}$ precede those of $\bar{F}$. Let S denote the linear span of all ordered monomials of the form

$$\bar{e}_1 \ldots \bar{e}_n \bar{f}_1 \ldots \bar{f}_m, \quad n \geq 0, \; m \geq 1,$$

with $\bar{e}_1 \leq \ldots \leq \bar{e}_n \in \bar{E}$, $\bar{f}_1 \leq \ldots \leq \bar{f}_m \in \bar{F}$. Since $\bar{f}_i \bar{e}_j = \bar{e}_j \bar{f}_i + ((\bar{f}_i, \bar{e}_j))$ and $((\bar{f}_i, \bar{e}_j)) \in J$, it follows that $S = J\underline{\underline{A}}$ is a two-sided ideal of $\underline{\underline{A}}$. It is now clear that the ordered E-monomials of the form $e_1 \ldots e_n$ with $e_1 \leq \ldots \leq e_n$ form an additive basis of the quotient ring $\underline{\underline{A}}/S$ and all it remains to prove is that $\underline{\underline{A}}/S$ is a universal envelope of $L = \underline{\underline{L}}(A)/J$. Let $\eta: L \to V$ be a Lie ring homomorphism of L to an associative ring V. Then η can be lifted to $\bar{\eta}: \underline{\underline{L}}(A) \to V$ with $\ker \bar{\eta} \supseteq J$. Since $\underline{\underline{A}}$ is the universal envelope for $\underline{\underline{L}}(A)$ we have a unique homomorphism $\bar{\theta}: \underline{\underline{A}} \to V$ with $\ker \bar{\theta} \supseteq S$. Thus there exists a unique homomorphism $\theta: \underline{\underline{A}}/S \to V$ extending η. This completes the proof of the theorem. □

A discussion with Yuri Bahturin is gratefully acknowledged.

Chapter II

Applications of Magnus Embedding

The Magnus embedding of F/R' into $\underline{\underline{M}}$ (Corollary I.1.10) provides a powerful tool in group theory. This was first demonstrated by Bachmuth & Hughes (1966). We illustrate by giving some applications. For some other applications and related embeddings we refer to Bachmuth (1965), Cohen (1967), Romanovskii (1972) & (1974), Remeslennikov (1973), Šmelkin (1964) and Dunwoody (1969).

1. SOME PROPERTIES OF F/R'

Let F be a non-cyclic free group with basis X and R a normal subgroup of F. Recall from Chapter I that $F/R' \cong \langle\psi(x);\ x \in X\rangle$, where

$$\psi(x) = \begin{bmatrix} xR & \lambda_x \\ 0 & 1R \end{bmatrix}$$

and, for $w \in F$,

$$\psi(w) = \begin{bmatrix} wR & \alpha_{12}(w) \\ 0 & 1R \end{bmatrix},$$

where

$$\alpha_{12}(w) = \sum_x \theta\ \partial_x(w)\lambda_x$$

and $\theta\colon \mathbb{Z}F \to \mathbb{Z}(F/R)$ is the linear extension of the natural projection $F \to F/R$ ($\partial_x(w)$ being the left Fox derivative of w with respect to x).

1.1 *THEOREM*. (Higman 1955). F/R' is always torsion-free.

Proof. Let $w \in F\backslash\{1\}$, $w \notin R'$, with $w^n \in R'$ for some $n > 1$. Since R/R' is free abelian, $w \notin R$. Set $S = \langle R, w\rangle$. Then S is free with $R \triangleleft S$ and S/R finite cyclic. Thus, without loss of generality, we may assume that $F = S$ and $F/R = \langle wR\rangle$ so that

$F' \leq R$. Now, $w^n \in R'$ implies that $\alpha_{12}(w^n) = 0$ and, in turn, $\theta\partial_x(w^n) = 0$ for all $x \in X$. This yields

$$\theta\{(1 + w + \ldots + w^{n-1})\partial_x(w)\} = 0$$

for all $x \in X$. Applying the unit augmentation $\varepsilon: \mathbb{Z}(F/R) \to \mathbb{Z}$ to both sides of the above yields $n\varepsilon\partial_x(w) = 0$ and consequently $\varepsilon\partial_x(w) = 0$ for all $x \in X$. Thus $\partial_x(w) \in \underline{\underline{f}}$ for all x and, in turn, $w \in F'$. Since $F' \leq R$, this gives $w \in R$ contrary to our choice. □

1.2 *THEOREM*. (Auslander & Lyndon 1955). If F/R' is abelian then $R = F$.

Proof. For any $x,y \in X$, $xy \equiv yx \pmod{R'}$ implies that $\alpha_{12}(xy) = \alpha_{12}(yx)$. This yields, in turn, $\theta(x)\lambda_y + \lambda_x = \theta(y)\lambda_x + \lambda_y$; $\theta(x) = 1$, $\theta(y) = 1$; $x \in R$, $y \in R$. Thus $F = R$. □

Theorem 1.2 admits an extension to F/R' being nilpotent due to B. H. Neumann (1962). We first establish an elementary lemma about group rings in general.

1.3 *LEMMA*. Let G be a group and $u \neq 0$ be an element of $\mathbb{Z}G$.

(i) If $(g-1)u = 0$ for all $g \in G$ then G is finite and $u = n \sum_{g\in G} g$ for some $n \in \mathbb{Z}$;

(ii) If $(g-1)(h-1)u = 0$ for all $g \in G$ then $(h-1)u = 0$.

Proof. Let $u = n_1g_1 + \ldots + n_tg_t$, $n_i \in \mathbb{Z}\backslash\{0\}$, $g_i \in G$ all distinct. Without loss of generality let $g_1 = 1$. Now $gu = u$ for all g implies that $g \in \{g_1 \ldots, g_t\}$ for all g and consequently $G = \{g_1,\ldots,g_t\}$ is finite. Further, $g_iu = u$ implies $n_1g_i = n_ig_i$ so that $n_1 = n_i$ for all i. Thus $u = n_1(1+g_2+\ldots+g_t) = n_1 \sum_{g\in G} g$. This proves (i). For the proof of (ii), put $v = (h-1)u$. If $v \neq 0$ then $(g-1)v = 0$ for all $g \in G$ implies by (i) that $v = n \sum_{g\in G} g$ for some $n \in \mathbb{Z}$. Applying the augmentation $\varepsilon: \mathbb{Z}G \to \mathbb{Z}$ to both sides yields $0 = n|G|$ and consequently $n = 0$. Thus $v = (h-1)u = 0$. □

1.4 *COROLLARY*. Let $G = F/R$ and $r \in R$.

(i) If $fR \in F/R$ is of infinite order then $[r,f] \in R'$ implies $r \in R'$;

(ii) If $[r,f] \in R'$ for all $f \in F$ and if $r \notin R'$ then $F/R = G$ is finite and $\alpha_{12}(r) = \sum_{x\in X} n_x \left(\sum_{g\in G} g\right)\lambda_x$, $n_x \in \mathbb{Z}$;

(iii) If $[r,f,g] \in R'$ for all $g \in F$ then $[r,f] \in R'$.

Proof. If $[r,f] \in R'$ then we have, in turn,

$$0 = \alpha_{12}[r,f];\ 0 = \theta(f-1)\ \alpha_{12}(r);\ 0 = \theta(f-1)\ \sum_{x\in X} \theta\partial_x(r)\lambda_x.$$

It follows that $\theta(f-1)\theta\partial_x(r) = 0$ for all x. Each of (i), (ii), (iii) now follows easily from Lemma 1.3. □

1.5 *THEOREM*. (B.H. Neumann 1962). If F/R' is an Engel group then $R = F$. In particular, if F/R' is nilpotent then $R = F$.

Proof. Let $r \in R$, $r \notin R'$. If, for some $x \in X$, $x \notin R$ then we set $S = \langle R,x\rangle$ so that S is free with S/R cyclic and S/R' an Engel group. Thus, for some $n \geq 1$, the left-normed commutator $[r,\underbrace{x,\ldots,x}_{n}] \in R'$ and, since S/R is cyclic, it follows that $[r,s_1,\ldots,s_n] \in R'$ for all $s_i \in S$. By Corollary 1.4(iii) it follows that $[r,s] \in R'$ for all $s \in S$. This yields S/R' abelian and $S = R$ by Theorem 1.2. This is contrary to the assumption that $x \notin R$. Thus $X \subseteq R$ and $F = R$ as required. □

1.6 *COROLLARY*. (Auslander & Lyndon 1955). If R,S are normal subgroups of F with $R' \leq S'$ then $R \leq S$. In particular, $R' = S'$ if and only if $R = S$.

Proof. Let $R' \leq S'$ with $R \not\leq S$. Then there exists $r \in R$ with $r \notin S$. Set $T = \langle S,r\rangle$. Then T is a free group with T/S cyclic. Now $R' \leq S'$ implies $\gamma_3(T) \leq S'$ so that T/S' is nilpotent (of class at most 2). Thus $T = S$ by Theorem 1.5. This is contrary to our choice of r. □

1.7 *REMARKS*. The normality requirement of R,S in Corollary 1.6 cannot be relaxed as the following example due to Dunwoody (1965) shows. Let $F = \langle x,y\rangle$, $S = \langle x,y^2,yxy^{-1}\rangle$, $R = \langle yx,y^2\rangle$. Thus $S \trianglelefteq F$, $R \ntrianglelefteq F$, $R \nleq S$. However, since $[yx,y^2] = [x,y^2]$, $R' \leq S'$.□

Let $\zeta(G)$ denote the centre of a group G and, for $k \geq 1$, let $\zeta_k(G)$ denote the k-th centre of G $(\zeta_1(G) = \zeta(G))$. Then $\zeta_k(G) = \{g \in G \mid [g,h_1,\ldots,h_k] = 1 \text{ for all } h_i \in G\}$. We next prove

1.8 *PROPOSITION*. (Auslander & Lyndon 1955). $\xi(F/R) \leq R/R'$.

Proof. Let $wR' \in \zeta(F/R')$ with $w \notin R$. Set $S = \langle R,w\rangle$. Then S is free with S/R cyclic. By hypothesis $[w,f] \in R'$ for all $f \in F$ and it follows that S/R' is abelian. By Theorem 1.2 we have $S = R$, contrary to our choice of w. □

1.9 *THEOREM*. (Auslander & Lyndon 1955). $\zeta(F/R')$ is non-trivial if and only if F/R is finite.

Proof. Let $wR' \in \zeta(F/R')$ with $w \notin R'$. Then, by Proposition 1.8, $w \in R$ and, by hypothesis, $[w,f] \in R'$ for all $f \in F$. By Corollary 1.4, F/R is finite and, in addition,

$$\alpha_{12}(w) = \sum_x n_x \Big(\sum_g g\Big)\lambda_x , \quad g \in G = F/R.$$

Conversely, let $F/R = G$ be finite. Then clearly the matrix

$$\begin{bmatrix} 1R & (\sum_g g)\lambda_x \\ 0 & 1R \end{bmatrix}, \quad x \in X,$$

lies in the centre of the ring M (see I.1.1). Thus it suffices to prove that $(\sum_g g)\lambda_x = \alpha_{12}(r)$ for some $r \in R$. Indeed, let $x^n \in R$ (n least) and let $h_1R,\ldots,h_mR$ be a set of coset representatives of F/R with respect to $\langle x\rangle R/R$. Then

$$\alpha_{12}(x^{nh_1}\ldots x^{nh_m}) = \sum_{i=1}^{m} \theta h_i \cdot \theta(1+x+\ldots+x^{n-1})\ \lambda_x = \Big(\sum_g g\Big)\lambda_x. \qquad \square$$

1.10 *REMARKS*. Let C/R' be the centre of F/R' with F/R finite. It follows from the above discussion that C/R' is a free abelian group with basis consisting of the matrices

$$\begin{bmatrix} 1R & (\sum_g g)\lambda_x \\ 0 & 1R \end{bmatrix}, \quad x \in X.$$

Thus rank $(C/R') = \text{rank}(F)$. By the Schreir formula, rank(R) is equal to $1 + |F/R|(\text{rank}(F)-1)$. It follows that if $|X| \geq 2$ then C/R' is a proper subgroup of R/R'. □

1.11 *THEOREM*. (Ojanguren 1968). If F/R is finite then C/R' is a direct summand of R/R'.

Proof. It clearly suffices to show that R/CR' is torsion free.

Let $r \in R$, $r \notin CR'$, with $r^n \in CR'$ for some $n \geq 2$. Then $[r^n,f] \in R'$ for all $f \in F$ and, in turn, $[r,f]^n \in R'$ for all $f \in F$; $[r,f] \in R'$ for all $f \in F$; $r \in CR'$, contrary to our choice of r. □

1.12 *PROPOSITION*. (Auslander & Lyndon 1955). Let $w \in F$ be such that $[r,w] \in R'$ for all $r \in R$. Then $w \in R$.

Proof. If $w \notin R$ we set $S = \langle R,w \rangle$. Then S is free with S/R cyclic and it follows that $R/R' \leq \zeta(S/R') < R/R'$, by Remark 1.10. □

1.13 *THEOREM*. (Gruenberg 1962). $\zeta_2(F/R') = \zeta_1(F/R')$.

Proof. Clearly we need only consider the case when F/R is finite. Let $wR' \in \zeta_2(F/R')$. We first prove that $w \in R$. Suppose $w \notin R$ and set $S = \langle R,w \rangle$. Then for all $r \in R$, $s_1,s_2 \in S$, we have $[r,s_1,s_2] \in R'$ and, by Corollary 1.4(iii), $[r,s] \in R'$ for all $r \in R$, $s \in S$. It follows that S/R' is abelian and hence $S = R$ by Theorem 1.2. Thus $wR' \in \zeta_2(F/R')$ implies that $w \in R$. Now $[w,f,g] \in R'$ for all $f,g \in F$ implies by Corollary 1.4(iii), that $[w,f] \in R'$ for all $f \in F$ and hence $wR' \in \zeta(F/R')$ as required. □

1.14 *THEOREM*. (Mal'cev 1960). Let F/R be torsion free. Then two elements of F/R' commute if and only if they either belong to R/R' or belong to the same cyclic subgroup of F/R'.

Proof. Let $[u,v] \in R'$ for some $uR',vR' \in F/R'$. If $u \in R$ and $v \notin R$ then, since F/R is torsion-free, by Corollary 1.4(i) it follows that $u \in R'$. Thus we may assume that $u \notin R$, $v \notin R$ and proceed to prove that $uR',vR' \in \langle wR' \rangle$ for some $w \in F$. Set $S = \langle R,u,v \rangle$. Then S/R is abelian and being torsion-free, has rank 1 or 2. If $\operatorname{rank}(S/R) = 2$ then $S/S' = S/R \oplus R/S'$ implies that u and v can be included in a free basis of S. In terms of this new basis for S, we have

$$\begin{bmatrix} uR & \lambda_u \\ 0 & 1R \end{bmatrix}\begin{bmatrix} vR & \lambda_v \\ 0 & 1R \end{bmatrix} = \begin{bmatrix} vR & \lambda_v \\ 0 & 1R \end{bmatrix}\begin{bmatrix} uR & \lambda_u \\ 0 & 1R \end{bmatrix}$$

and, as in the proof of Theorem 1.2, it follows that $u,v \in R$ contrary to our choice. Thus $\operatorname{rank}(S/R) = 1$. Consequently, $u \equiv w^i \pmod R$, $v \equiv w^j \pmod R$ for some $w \in S$, $w \notin R$. Set

$d = \text{g.c.d.}\{i,j\}$ so that $\text{g.c.d.}\{i/d, j/d\} = 1$. Then $u^{j/d} \equiv w^{ij/d} \equiv v^{i/d} \pmod{R}$ and it follows that $u^{j/d} = v^{i/d} r$ for some $r \in R$. Now, $[u,v] \in R'$ implies $[r,u] \in R'$; thus, since S/R is torsion-free, $r \in R'$ by Corollary 1.4(i). It follows that $u^{j/d} \equiv v^{i/d} \pmod{R'}$. Choose $p, q \in \mathbb{Z}$ with $1 = p(i/d) + q(j/d)$. Then $u = u^{p(i/d)} u^{q(j/d)} = u^{p(i/d)} v^{q(i/d)}$ $\equiv (u^p v^q)^{i/d} \pmod{R'}$. Similarly, $v \equiv (u^p v^q)^{j/d} \pmod{R'}$. The result follows with $w = u^p v^q$. □

1.15 *THEOREM*. (Baumslag & Gruenberg 1967). Every locally nilpotent subgroup of F/R' is abelian.

Proof. Let N/R' be a locally nilpotent subgroup of F/R'. Let $uR', vR' \in N/R'$ with $u, v \notin R'$. Set $H = \langle u,v \rangle$. By Hypothesis, $\gamma_n(H) \leq R'$ for some $n \geq 2$. The proof consists in showing that $H' \leq R'$. If $u \in R$, $v \in R$ then there is nothing to prove. So we may assume $u \notin R$.

Case I. (HR/R torsion-free).

Suppose $n \geq 3$ and choose $w \in \gamma_{n-1}(H)$, $w \notin R'$. Then $[w,u] \in \gamma_n(H) \leq R'$ and it follows by Theorem 1.14 that $w \equiv f^i \pmod{R'}$, $u \equiv f^j \pmod{R'}$ for some $f \in F$ and $i,j \in \mathbb{Z}$. Thus $w^j \equiv u^i \pmod{R'}$ and, since $[w,v] \in \gamma_n(H) \leq R'$, $[w^j,v] \in R'$. It follows that $[u^i,v] \in R'$ which yields $[u,v]^i \in \gamma_3(H)R'$. By induction we conclude that $[u,v]^{i^{(n-1)}} \in \gamma_n(H)R' \leq R'$. Since R/R' is torsion-free (Theorem 1.1), it follows that $[u,v] \in R'$.

Case II. (HR/R not torsion-free).

Let $z \in H$ with $z^k \in R$ for some $k > 0$. Then the left-normed commutator $[z^k, h_1, \ldots, h_{n-1}] \in \gamma_n(H)R' \leq R'$ for all $h_i \in H$ and so $[z^k, h_1^*, \ldots, h_{n-1}^*] \in R'$ for all $h_i^* \in HR$. It follows that $z^k R' \in \zeta_{n-1}(HR/R') = \zeta(HR/R')$, by Theorem 1.13. Thus either $H \leq R$ which is a trivial case, or HR/R is finite (Theorem 1.9). This yields $u^i \in R$, $v^j \in R$ for some $i,j \geq 1$. Thus $[u^i,v^j] \in R'$ and, in turn, as before, $[u,v]^{ij} \in \gamma_3(H)R'$; $[u,v]^{(ij)^{n-2}} \in \gamma_n(H)R' \leq R'$; $[u,v] \in R'$. □

Before closing this section we prove the following generalization of a result due to Birman (1973) about the endomorphisms of free groups.

1.16 *THEOREM*. (Krasnikov 1978). Let $F = \langle x_1,\ldots,x_n \rangle$, $n \geq 2$, be a free group and $R \trianglelefteq F$. Then $\{w_1R',\ldots,w_nR'\}$, $w_i \in F$, generates F/R' if and only if the $n\times n$ matrix $[\theta\partial_{x_j}(w_i)]_{n\times n}$ is left-invertible over $\mathbb{Z}(F/R)$.

Proof. For $z \in \mathbb{Z}F$ we denote by $z[\underset{\sim}{w}]$ the evaluation $z\big|_{x_1=w_1,\ldots,x_n=w_n}$. We first prove the following chain rule due to Fox (1953):

$$\partial_{x_j}(f[\underset{\sim}{w}]) = \sum_{k=1}^{n} (\partial_{x_k}(f))[\underset{\sim}{w}]\cdot\partial_{x_j}(w_k) \tag{1}$$

for all $f \in F$.

We prove (1) by induction on the length of f. If $f = x_m^{\varepsilon_m}$, $\varepsilon_m = \pm 1$, then

$$\begin{aligned}\sum_{k=1}^{n} (\partial_{x_k}(f))[\underset{\sim}{w}]\cdot\partial_{x_j}(w_k) &= \partial_{x_m}(x_m^{\varepsilon_m})[\underset{\sim}{w}]\cdot\partial_{x_j}(w_m)\\ &= \varepsilon_m w_m^{(\varepsilon_m-1)/2}\,\partial_{x_j}(w_m)\\ &= \partial_{x_j}(w_m^{\varepsilon_m}) = \partial_{x_j}(f[\underset{\sim}{w}]).\end{aligned}$$

Let $f = gx_m^{\varepsilon_m}$ and assume that (1) holds for g. Then $f[\underset{\sim}{w}] = g[\underset{\sim}{w}]w_m^{\varepsilon_m}$ and so

$$\begin{aligned}\partial_{x_j}(f[\underset{\sim}{w}]) &= \partial_{x_j}(g[\underset{\sim}{w}]) + g[\underset{\sim}{w}]\cdot\partial_{x_j}(w_m^{\varepsilon_m})\\ &= \sum_{k=1}^{n} \partial_{x_k}(g)[\underset{\sim}{w}]\cdot\partial_{x_j}(w_k) + g[\underset{\sim}{w}]\cdot\partial_{x_j}(w_m^{\varepsilon_m}).\end{aligned} \tag{2}$$

Since $f = gx_m^{\varepsilon_m}$ we have

$$\partial_{x_k}(f) = \partial_{x_k}(g) + g\,\varepsilon_m\, x_m^{(\varepsilon_m-1)/2}\,\delta_{mk} \tag{3}$$

where

$$\delta_{ij} = \begin{cases} 1 & \text{if } i = j \\ 0 & \text{if } i \neq j. \end{cases}$$

Evaluating (3) at $\underset{\sim}{w}$ yields

$$\partial_{x_k}(f)[\underset{\sim}{w}] = \partial_{x_k}(g)[\underset{\sim}{w}] + g[\underset{\sim}{w}]\,\varepsilon_m\, w_m^{(\varepsilon_m-1)/2}\,\delta_{mk}. \tag{4}$$

Substituting the value of $\partial_{x_k}(g)[\underset{\sim}{w}]$ from (4) in (2) yields

$$\begin{aligned}
\partial_{x_j}(f[\underset{\sim}{w}]) &= \sum_{k=1}^{n} \partial_{x_k}(f)[\underset{\sim}{w}]\cdot\partial_{x_j}(w_k) \\
&\quad - \sum_{k=1}^{n} g[\underset{\sim}{w}]\varepsilon_m\, w_m^{(\varepsilon_m-1)/2}\,\delta_{mk}\partial_{x_j}(w_k) \\
&\quad + g[\underset{\sim}{w}]\cdot\partial_{x_j}(w_m^{\varepsilon_m}) \\
&= \sum_{k=1}^{n} \partial_{x_k}(f)[\underset{\sim}{w}]\cdot\partial_{x_j}(w_k) - g[\underset{\sim}{w}]\varepsilon_m w_m w_m^{(\varepsilon_m-1)/2}\,\partial_{x_j}(w_m) \\
&\quad + g[\underset{\sim}{w}]\cdot\partial_{x_j}(w_m^{\varepsilon_m}) \\
&= \sum_{k=1}^{n} \partial_{x_k}(f)[\underset{\sim}{w}]\cdot\partial_{x_j}(w_k)
\end{aligned}$$

This completes the proof of (1).

For the necessity part of the theorem let us assume that we have $F = \langle w_1,\ldots,w_n\rangle R'$. Then there exist $f_1,\ldots,f_n \in F$ and $r_1,\ldots,r_n \in R'$ such that $x_1 = f_1[\underset{\sim}{w}]r_1,\ldots,x_n = f_n[\underset{\sim}{w}]r_n$. Now

$$\begin{aligned}
\delta_{ij} = \partial_{x_j}(x_i) &= \partial_{x_j}(f_i[\underset{\sim}{w}]\ r_i) \\
&\equiv \partial_{x_j}(f_i[\underset{\sim}{w}])\ (\mathrm{mod}\ \underset{=}{r}) \\
&\equiv \sum_{k=1}^{n} \partial_{x_k}(f_i)[\underset{\sim}{w}]\cdot\partial_{x_j}(w_k)\ (\mathrm{mod}\ \underset{=}{r})
\end{aligned}$$

(since $\partial_{x_j}(r_i) \equiv 0\ (\mathrm{mod}\ \underset{=}{r})$, $\underset{=}{r} = \mathbb{Z}F(R-1)$).

Thus with $\theta\colon \mathbb{Z}F \to \mathbb{Z}(F/R)$,

$$\delta_{ij} = \sum_{k=1}^{n} \theta\partial_{x_k}(f_i)[\underset{\sim}{w}]\cdot\theta\partial_{x_j}(w_k)$$

and it follows that $I = ED$, where I is the $n\times n$ identity matrix,

$$E = [\theta\partial_{x_j}(f_i)[\underset{\sim}{w}]]_{n\times n} \quad\text{and}\quad D = [\theta\partial_{x_j}(w_i)]_{n\times n}$$

are $n\times n$ matrices and E is the required left-inverse of D.

For the sufficiency part of the theorem let $E = [\theta u_{ij}]_{n\times n}$ be a left-inverse of $[\theta\partial_{x_j}(w_i)]_{n\times n}$ over $\mathbb{Z}(F/R)$. Since $w_i - 1 = \sum_{k=1}^{n} \partial_{x_k}(w_i)(x_k - 1)$ (Theorem I.2.2(c)), it follows that

$$\begin{bmatrix} w_1-1 \\ \cdot \\ \cdot \\ \cdot \\ w_n-1 \end{bmatrix} = [\partial_{x_j}(w_i)]_{n\times n} \begin{bmatrix} x_1-1 \\ \cdot \\ \cdot \\ \cdot \\ x_n-1 \end{bmatrix}$$

so that

$$\begin{bmatrix} \theta(x_1-1) \\ \cdot \\ \cdot \\ \cdot \\ \theta(x_n-1) \end{bmatrix} = E \begin{bmatrix} \theta(w_1-1) \\ \cdot \\ \cdot \\ \cdot \\ \theta(w_n-1) \end{bmatrix}$$

and, in particular, modulo $\underline{\underline{r}}$,

$$x_i - 1 \in \text{left-ideal}_{\mathbb{Z}F}\{w_1-1,\ldots,w_n-1\},\ i = 1,\ldots,n.$$

Let $S = \langle w_1,\ldots,w_n\rangle$, $G = F/R$, $H = SR/R$. Then it follows that $\mathbb{Z}G(G-1) = \mathbb{Z}G(H-1)$. Since $G \cap (1+\mathbb{Z}G(H-1)) = H$, it follows that $F = \langle w_1,\ldots,w_n\rangle R$ which is a first step towards proving $F = \langle w_1,\ldots,w_n\rangle R'$. Using $F = \langle w_1,\ldots,w_n\rangle R$ we first express u_{ij} as

$$u_{ij} \equiv v_{ij}[\underset{\sim}{w}] \pmod{\underline{\underline{r}}}$$

where $v_{ij} \in \mathbb{Z}F$. Next we find $z_i \in \mathbb{Z}F$, $i = 1,\ldots,n$, such that

$$\partial_{x_j}(z_i) = v_{ij}$$

[for instance by Theorem I.2.2, $z_i = \sum_{j=1}^{n} v_{ij}(x_j-1)$ has the property that $\partial_{x_j}(z_i) = v_{ij}$]. It follows that there exist $z_1,\ldots,z_n \in \mathbb{Z}F$ such that

$$E = [\theta\partial_{x_j}(z_i)[\underset{\sim}{w}]]_{n\times n} \text{ is a left-inverse of}$$

$$D = [\theta\partial_{x_j}(w_i)]_{n\times n}\ . \qquad (5)$$

Since $z_i - \varepsilon(z_i) = \sum_{j=1}^{n} \partial_{x_j}(z_i)(x_j-1)$, $\varepsilon\colon \mathbb{Z}F \to \mathbb{Z}$, (Theorem I.2.2(c)), we have $z_i[\underset{\sim}{w}] - \varepsilon(z_i) = \sum_{j=1}^{n} \partial_{x_j}(z_i[\underset{\sim}{w}](w_j-1)$ for all i

and consequently

$$\begin{bmatrix} \theta z_1[\underset{\sim}{w}]-\varepsilon(z_1) \\ \vdots \\ \theta z_n[\underset{\sim}{w}]-\varepsilon(z_n) \end{bmatrix} = E \begin{bmatrix} \theta(w_1-1) \\ \vdots \\ \theta(w_n-1) \end{bmatrix} = \begin{bmatrix} \theta(x_1-1) \\ \vdots \\ \theta(x_n-1) \end{bmatrix}.$$

Thus, for $i = 1,\ldots,n$, we have

$$z_i[\underset{\sim}{w}]-\varepsilon(z_i) \equiv x_i-1 \pmod{\underline{r}}. \tag{6}$$

We now prove that it is possible to choose $z_1,\ldots,z_n$ such that each $z_i \in F$ and (5) holds. Assume on the contrary that $z_i \notin F$ for some i. Then we may write z_i as

$$z = z_i = \varepsilon_1 f_1 + \ldots + \varepsilon_t f_t, \tag{7}$$

where $\varepsilon_j = \pm 1$, $f_i \in F$, $t \geq 2$ is least possible. Then by (6),

$$z[\underset{\sim}{w}]-\varepsilon(z) = \varepsilon_1 f_1[\underset{\sim}{w}] + \ldots + \varepsilon_t f_t[\underset{\sim}{w}]-\varepsilon(z) \equiv x_i-1 \pmod{\underline{r}}.$$

Thus we must have one of the following two cases:

Case I. $\varepsilon_a f_a[\underset{\sim}{w}]+\varepsilon_b f_b[\underset{\sim}{w}] \equiv 0 \pmod{\underline{r}}$ for some $1 \leq a < b \leq t$,

Case II. $f_a[\underset{\sim}{w}] \equiv 1 \pmod{\underline{r}}$ for some $1 \leq a \leq t$.

In Case I, $\varepsilon_a+\varepsilon_b = 0$ and $f_a[\underset{\sim}{w}] = f_b[\underset{\sim}{w}]$, $f_a \neq f_b$. We choose $g = f_a f_b^{-1}$ so that $f_a = gf_b$ with $g[\underset{\sim}{w}] \equiv 1 \pmod{\underline{r}}$. Then

$$\partial_{x_j}(f_a) = \partial_{x_j}(g) + g\partial_{x_j}(f_b)$$

gives

$$\partial_{x_j}(g)[\underset{\sim}{w}] \equiv \partial_{x_j}(f_a)[\underset{\sim}{w}] - \partial_{x_j}(f_b)[\underset{\sim}{w}] \pmod{\underline{r}}$$

for all $j = 1,\ldots,n$. Consequently,

$$\partial_{x_j}(\varepsilon_a g)[\underset{\sim}{w}] \equiv \partial_{x_j}(\varepsilon_a f_a+\varepsilon_b f_b)[\underset{\sim}{w}] \pmod{\underline{r}}.$$

Let z_i' be obtained from z_i (given by (7)) upon replacing $\varepsilon_a f_a+\varepsilon_b f_b$ by $\varepsilon_a g$. Then $z_1,\ldots,z_i',\ldots,z_n$ satisfy (5) and z_i' has fewer components contrary to the minimality of t. In Case II, we may assume, without loss of generality, that $\varepsilon_1 f_1[\underset{\sim}{w}] \equiv x_i \pmod{\underline{r}}$ and $\varepsilon_2 f_2[\underset{\sim}{w}] \equiv -1 \pmod{\underline{r}}$ so that $\varepsilon_1 = 1$, $\varepsilon_2 = -1$ and $f_2[\underset{\sim}{w}] \equiv 1 \pmod{\underline{r}}$. Put $g = f_2^{-1} f_1$. Then

$\partial_{x_j}(g) = \partial_{x_j}(f_2^{-1}) + f_2^{-1}\partial_{x_j}(f_1) = -f_2^{-1}\partial_{x_j}(f_2) + f_2^{-1}\partial_{x_j}(f_1)$ yields $\partial_{x_j}(g)[\underset{\sim}{w}] \equiv \partial_{x_j}(f_1)[\underset{\sim}{w}] - \partial_{x_j}(f_2)[\underset{\sim}{w}]$ and, in turn,

$$\partial_{x_j}(\varepsilon_1 g)[\underset{\sim}{w}] \equiv \partial_{x_j}(\varepsilon_1 f_1 + \varepsilon_2 f_2)[\underset{\sim}{w}] \pmod{\underset{=}{\mathfrak{r}}}$$

for all $j = 1,\dots,n$. Thus as in the Case I replacing $\varepsilon_1 f_1 + \varepsilon_2 f_2$ by $\varepsilon_1 g$ in (7) replaces z_i by z_i' with fewer components contrary to the minimality. This proves that we can choose $z_1,\dots,z_n$ satisfying (5) such that $z_i \in F$. We set $u_i = z_i[\underset{\sim}{w}]$, $i = 1,\dots,n$. Then using (1) we have

$$\begin{aligned}\partial_{x_j}(u_i) &= \partial_{x_j}(z_i[\underset{\sim}{w}]) \\ &= \sum_{k=1}^{n} \partial_{x_k}(z_i)[\underset{\sim}{w}]\cdot\partial_{x_j}(w_k)\end{aligned}$$

which yields

$$[\theta\partial_{x_j}(u_i)]_{n\times n} = ED = I = [\partial_{x_j}(x_i)]_{n\times n}$$

so that $\theta\partial_{x_j}(x_i^{-1}u_i) = 0$ for all $j = 1,\dots,n$. By Theorem I.2.4 it follows that $x_i^{-1}u_i \in R'$ for all i and consequently $z_i[\underset{\sim}{w}] = u_i \equiv x_i \pmod{R'}$. This proves that $F = \langle w_1,\dots,w_n\rangle R'$ as desired. □

1.17 REMARKS. One interesting result not covered here is that if F/R is hopfian then so is F/R' (Dunwoody 1971) [see also Krasnikov 1978]. □

2. SOME ALGORITHMIC PROBLEMS IN F/R'

The Magnus embedding of F/R' is particularly suitable for transferring certain decision problems of F/R' to those of F/R. If F/R has the solvable word problem (i.e., there is an algorithm which decides in a finite number of steps whether or not any given element of F/R is the identity element) then the group ring $\mathbb{Z}(F/R)$ has the solvable word problem (i.e., there is an algorithm to decide if any given element of $\mathbb{Z}(F/R)$ is the zero element). It follows then that the $\mathbb{Z}(F/R)$-module $\Omega = \mathbb{Z}(F/R)[\lambda_x \mid x \in X]$ has the solvable word problem and, in turn, the ring $M = \begin{bmatrix}\mathbb{Z}(F/R) & \Omega \\ 0 & \mathbb{Z}\end{bmatrix}$ has the solvable word problem. In particular, the group F/R' (which by Corollary I.1.10 is embedded in M) has the solvable word problem. An iteration of this observation yields the following important result.

2.1 *THEOREM*. If F/R has the solvable word problem then $F/\delta_n(R)$ has the solvable word problem for all $n \geq 1$, where $\delta_n(R)$ denotes the n-th term of the derived series of R (i.e., $\delta_n(R) = [\delta_{n-1}(R), \delta_{n-1}(R)]$, $n \geq 2$). □

Since the free abelian group F/F' has the solvable word problem, Theorem 2.1 yields,

2.2 *COROLLARY*. Free solvable groups have the solvable word problem. □

A group G is said to have the solvable power problem if for given g and h in G, it can be decided whether or not there exists an integer n such that $g = h^n$.

2.3 *THEOREM*. If F/R is a torsion-free group with the solvable power problem then F/R' has the solvable power problem.

Proof. Let $u, v \in F/R'$ and let

$$\psi(u) = \begin{bmatrix} uR & \alpha_{12}(u) \\ 0 & 1R \end{bmatrix} \qquad \psi(v) = \begin{bmatrix} vR & \alpha_{12}(v) \\ 0 & 1R \end{bmatrix}$$

be the matrix images of u, v under the Magnus embedding of F/R' as given by I.1.10. Using the solvable power problem of F/R we decide whether or not $uR \in \langle vR \rangle$. If $uR \notin \langle vR \rangle$ then clearly $\psi(u) \notin \langle \psi(v) \rangle$. If $uR \in \langle vR \rangle$ and $uR = (vR)^n$, say, then since F/R is torsion-free $uR' \in \langle vR' \rangle$ if and only if $uR' = v^n R'$. Thus we need only to decide the validity of the equation $\alpha_{12}(u) = \alpha_{12}(v^n)$ which is clearly possible since $\mathbb{Z}(F/R)$ has the solvable word problem. □

A group G is said to have the solvable conjugacy problem if for given $g, h \in G$ it can be decided whether or not $g = u^{-1}hu$ for some $u \in G$. Solvability of the conjugacy problem is perhaps the most significant property which is carried over to F/R from F/R'.

2.4 *THEOREM*. (Remeslennikov & Sokolov 1970, C.K. Gupta 1982). Let F/R be a finitely generated recursively presented group with the solvable conjugacy problem and the solvable power problem. Then F/R' has the solvable conjugacy problem.

Proof. Let T be a right-transversal of F with respect to R. Let $\psi(ra)$, $\psi(sb)$ be two given elements of F/R' where $r,s \in R$, $a,b \in T$ and $\psi\colon F/R' \to M$ is the Magnus embedding. Then we wish to decide the existence of an element $\psi(tc)$, $t \in R$, $c \in T$ such that

$$\psi(ra) = (\psi(tc))^{-1}\psi(sb)\psi(tc).$$

It is clearly necessary that aR and bR are conjugate so we may assume that there exists $c \in T$ with

$$aR = (cR)^{-1}\, bR\, cR.$$

Since $\psi(c)^{-1}\psi(sb)\psi(c) = \psi(s'a)$ for some $s' \in R$, replacing $\psi(sb)$ by $\psi(s'a)$, if necessary, we may assume that the given elements are

$$\psi(ra) = \begin{bmatrix} \theta a & \alpha_{12}(ra) \\ 0 & \theta 1 \end{bmatrix} \quad \text{and} \quad \psi(sa) = \begin{bmatrix} \theta a & \alpha_{12}(sa) \\ 0 & \theta 1 \end{bmatrix},$$

where $\theta\colon \mathbb{Z}F \to \mathbb{Z}(F/R)$ is the natural projection. Since

$$\begin{bmatrix} \theta f & 0 \\ 0 & \theta 1 \end{bmatrix}\begin{bmatrix} \theta 1 & \lambda_x \\ 0 & \theta 1 \end{bmatrix}\begin{bmatrix} \theta f & 0 \\ 0 & \theta 1 \end{bmatrix}^{-1} = \begin{bmatrix} \theta 1 & \theta f\lambda_x \\ 0 & \theta 1 \end{bmatrix}$$

and

$$\begin{bmatrix} \theta 1 & \theta f_x\lambda_x \\ 0 & \theta 1 \end{bmatrix}\begin{bmatrix} \theta 1 & \theta f_y\lambda_y \\ 0 & \theta 1 \end{bmatrix} = \begin{bmatrix} \theta 1 & \theta f_x\lambda_x + \theta f_y\lambda_y \\ 0 & \theta 1 \end{bmatrix},$$

it is easily verified that the multiplicative group

$$G = \begin{bmatrix} \theta F & \sum\limits_{x} \theta F\lambda_x \\ 0 & \theta 1 \end{bmatrix}$$

is isomorphic to the restricted wreath product $A \text{ wr } B$ where

$$A = \left\langle \begin{bmatrix} \theta 1 & \lambda_x \\ 0 & \theta 1 \end{bmatrix},\ x \in X \right\rangle \cong F/F'$$

and

$$B = \begin{bmatrix} \theta F & 0 \\ 0 & \theta 1 \end{bmatrix} \cong F/R.$$

Since B has the solvable conjugacy problem and the solvable power problem, it follows by a result of Matthews (1966) that G ($\cong$ A wr B) has the solvable conjugacy problem (or, see Remark 2.5). Since F/R' is embedded in G, there exists an element

$$\begin{bmatrix} \theta c & z \\ 0 & \theta 1 \end{bmatrix} \in G, \; c \in T, \; z \in \Omega = \sum_x \mathbb{Z}(F/R)\lambda_x ,$$

such that

$$\begin{bmatrix} \theta c & z \\ 0 & \theta 1 \end{bmatrix} \begin{bmatrix} \theta a & \alpha_{12}(ra) \\ 0 & \theta 1 \end{bmatrix} = \begin{bmatrix} \theta a & \alpha_{12}(sa) \\ 0 & \theta 1 \end{bmatrix} \begin{bmatrix} \theta c & z \\ 0 & \theta 1 \end{bmatrix} \tag{8}$$

[If no such element of G exists then clearly $\psi(ra)$ and $\psi(sa)$ are not conjugate.]

The proof consists in showing that z can be effectively replaced by $\alpha_{12}(cq)$ for some $q \in R$.

From (8), matrix multiplication yields

$$ca \equiv ac \pmod R \tag{9}$$

and

$$\theta(a-1)z = \theta c \cdot \alpha_{12}(ra) - \alpha_{12}(sa). \tag{10}$$

If $a = 1$, we simply choose $c \in T$ such that $\theta c \cdot \alpha_{12}(ra) = \alpha_{12}(sa)$ (which is clearly decidable) and it follows that $z = \alpha_{12}(c)$ is a required solution of (8). If $a \neq 1$, then (9) yields

$$ca = qac$$

for some $q \in R$ and it follows that

$$\alpha_{12}(ca) = \alpha_{12}(qac)$$

which gives (multiplying the corresponding matrices)

$$\theta(c-1) \cdot \alpha_{12}(a) = \theta(a-1) \cdot \alpha_{12}(c) + \alpha_{12}(q). \tag{11}$$

Next,

$$\begin{aligned}\theta c\cdot\alpha_{12}(ra)-\alpha_{12}(sa) &= \theta c\cdot(\alpha_{12}(a)+\alpha_{12}(r))-(\alpha_{12}(a)+\alpha_{12}(s))\\ &= \theta(c-1)\cdot\alpha_{12}(a) + \theta c\cdot\alpha_{12}(r) - \alpha_{12}(s)\\ &= \theta(a-1)\cdot\alpha_{12}(c) + \alpha_{12}(q) + \theta c\cdot\alpha_{12}(r)\\ &\qquad - \alpha_{12}(s), \text{ by (11). Thus}\end{aligned}$$

$$\theta c\cdot\alpha_{12}(ra)-\alpha_{12}(sa) = \theta(a-1)\alpha_{12}(c) + \alpha_{12}(q\cdot crc^{-1}\cdot s^{-1}), \tag{12}$$

which by (10) gives

$$\theta(a-1)z^* = \alpha_{12}(q^*), \tag{13}$$

where

$$z^* = \theta z - \alpha_{12}(c), \qquad q^* = q\cdot crc^{-1}\cdot s^{-1} \in R.$$

It remains to show that z^* in (13) can be effectively replaced by $\alpha_{12}(t^*)$ for some $t^* \in R$. For, then

$$\theta(a-1)\cdot\alpha_{12}(t^*) = \alpha_{12}(q^*)$$

implies

$$\theta(a-1)\cdot\alpha_{12}(t^*) + \theta(a-1)\cdot\alpha_{12}(c) = \theta(a-1)\cdot\alpha_{12}(c) + \alpha_{12}(q^*)$$

and, in turn,

$$\theta(a-1)\cdot\alpha_{12}(t^*c) = \theta(a-1)\cdot\alpha_{12}(c) + \alpha_{12}(q\cdot crc^{-1}\cdot s^{-1});$$

$$\theta(a-1)\cdot\alpha_{12}(t^*c) = \theta c\cdot\alpha_{12}(ra) - \alpha_{12}(sa), \text{ by (12)},$$

which by comparison with (10) yields

$$\psi(t^*c)\psi(ra) = \psi(sa)\psi(t^*c)$$

as required.

In (13), let $\alpha_{12}(q^*) = \sum_x \theta g_x\lambda_x$ and $z^* = \sum_x \theta z_x^*\lambda_x$. Then (13) yields

$$\theta(a-1)\cdot\theta z_x^* = \theta g_x \quad \text{for all } x \in X$$

and consequently, in turn,

$$\theta((a-1)\cdot z_x^*(x-1)) = \theta(g_x(x-1)) \quad \text{for all } x;$$

$$\theta(a-1)\cdot\theta\sum_x z_x^*(x-1) = \theta\sum_x g_x(x-1);$$

$$\theta(a-1)\cdot\theta\sum_x z_x^*(x-1) = 0, \text{ by Theorem I.2.5},$$

since $q^* \in R$.

If aR is of infinite order then, by Lemma 1.3, $\theta \sum_x z_x^*(x-1) = 0$ and, by Theorem I.2.5, $z^* = \alpha_{12}(t^*)$ for some $t^* \in R$.

If aR is of finite order n, say, then

$$\theta \sum_x z_x^*(x-1) = \theta p(a) \cdot h \tag{14}$$

for some effectively determined $h \in \mathbb{Z}(F/R)$ and $p(a) = 1 + a + \ldots + a^{n-1}$. Since the left hand side of (14) lies in the augmentation ideal of $\mathbb{Z}(F/R)$, it follows that

$$h = \theta \sum_x h_x(x-1), \quad h_x \in \mathbb{Z}F$$

and (14) becomes

$$\theta \sum_x (z_x^* - p(a)h_x)(x-1) = 0.$$

Again, by Theorem I.2.5, it follows that for some $t^* \in R$,

$$\theta \sum_x (z_x^* - p(a)h_x)\lambda_x = \alpha_{12}(t^*).$$

Thus, with $h^* = \sum_x h_x \lambda_x$,

$$z^* - \theta(p(a)h^*) = \alpha_{12}(t^*)$$

and consequently

$$\theta(a-1) \cdot z^* = \theta(a-1) \cdot \alpha_{12}(t^*)$$

as was to be proved. □

2.5 *REMARKS*. To solve the conjugacy problem for G directly, it clearly suffices to solve it for the group $\begin{bmatrix} G & \mathbb{Z}G \\ 0 & 1 \end{bmatrix}$ assuming that G has the solvable conjugacy and power problems. Thus given $u = \begin{bmatrix} a & \alpha \\ 0 & 1 \end{bmatrix}$ and $v = \begin{bmatrix} b & \beta \\ 0 & 1 \end{bmatrix}$, $a, b \in G$, $\alpha, \beta \in \mathbb{Z}G$, we wish to decide the existence of some $w = \begin{bmatrix} c & \gamma \\ 0 & 1 \end{bmatrix}$, $c \in G$, $\gamma \in \mathbb{Z}G$, such that $w^{-1}uw = v$. Using the solvability of the conjugacy problem for G we first find $d \in G$ such that $d^{-1}ad = b$ (otherwise u and v are not conjugate). With $z = \begin{bmatrix} d & 0 \\ 0 & 1 \end{bmatrix}$, we have

$v' = z\,v\,z^{-1} = \begin{bmatrix} a & d\beta \\ 0 & 1 \end{bmatrix}$. Since u and v are conjugate if and only if u and v' are conjugate, without loss of generality we may assume that $v = \begin{bmatrix} a & \beta \\ 0 & 1 \end{bmatrix}$. Now $uw = wv$ if and only if (i) $ac = ca$ and (ii) $a\gamma+\alpha = c\beta+\gamma$. Thus we seek $c \in G$, $\gamma \in \mathbb{Z}G$ such that c commutes with a and $(a-1)\gamma = c\beta-\alpha$. We use the solvability of the power problem of G to determine whether a has finite or infinite order. If a has finite order, say p, then we can determine the existence of $c \in G$ such that $c\beta^*-\alpha^* = 0$, where $\alpha^* = (1+a+\ldots+a^{p-1})\alpha$ and $\beta^* = (1+a+\ldots+a^{p-1})\beta$, express $c\beta-\alpha$ as $(a-1)\gamma'$ and choose $\gamma = \gamma'$. If a is of infinite order then we need to determine the existence of $c \in G$ such that $ca = ac$ and $c\beta-\alpha$ is of the form $(a-1)\gamma'$ so that choosing $\gamma = \gamma'$ yields the required $w = \begin{bmatrix} c & \gamma \\ 0 & 1 \end{bmatrix}$. (By Matthews' theorem the existence of c is decidable. A direct proof should be possible.) □

2.6 *REMARKS*. Remeslennikov & Sokolov (1970) proved Theorem 2.4 with the additional assumption that F/R be torsion-free. The present version and its proof are due to C.K. Gupta (1982). □

An immediate consequence of Theorems 2.3 and 2.4 is the following result.

2.7 *COROLLARY*. (Kargapolov & Remeslennikov 1966). Free solvable groups have the solvable conjugacy problem. □

3. FAITHFULNESS OF RELATION MODULES

Let $1 \to G \to F \to R \to 1$ be a free presentation of a group G. Then R/R' can be regarded as a right G-module via conjugation in F, i.e., if $rR' \in R/R'$, $g = fR \in F/R$, then

$$rR'\cdot g = (f^{-1}rf)R' = r^fR'.$$

This module is called a <u>relation module</u> for G. We shall need to obtain precise information about the structure of $R/R' \otimes_{\mathbb{Z}} \mathbb{Q}$ when F/R is finite. Let us first compute the action of F/R when it is finite cyclic of order $e \geq 2$ and $F = \langle x_1,\ldots,x_m \rangle$, $m \geq 2$. Since F/F' is free abelian of rank m and R/F' is of index

e in F/F', we can choose a basis $\{y_1,\ldots,y_m\}$ for F such that $y_1^e, y_2,\ldots,y_m \in R$. Now R is generated by $y_1^e, y_1^{-i}y_2y_1^i,\ldots,y_1^{-i}y_my_1^i$, $i \in \{0,\ldots,e-1\}$, and the number of these generators adds up to $1+(m-1)e$ which coincides with the rank of R by the Schreier formula. Thus R is freely generated by these elements. The action of F/R on these generators yields the following result.

3.1 *PROPOSITION.* (Ojanguren 1968). If rank $F = m \geq 2$ and if $G = F/R$ is finite cyclic then $R/R' = \mathbb{Z} \oplus \underbrace{\mathbb{Z}G \oplus \ldots \oplus \mathbb{Z}G}_{\leftarrow m-1 \rightarrow}$. □

Let $T(y_1)$ be the matrix of the linear transformation induced by the action of y_1 on $R/R' \otimes_{\mathbb{Z}} Q$, where Q is the field of rationals. Then

$$T(y_1) = I \oplus \underbrace{E \oplus \ldots \oplus E}_{\leftarrow m-1 \rightarrow} \tag{15}$$

where I is the 1×1 identity matrix and E is the $e\times e$ permutation matrix given by

$$E = \begin{bmatrix} 0 & 1 & 0 & \ldots & 0 \\ 0 & 0 & 1 & \ldots & 0 \\ & & & & \\ & & & & 1 \\ 1 & 0 & 0 & \ldots & 0 \end{bmatrix}.$$

It follows that the trace of $T(y_1^i)$ is 1 for all $i = 1,\ldots,e-1$.

We can now prove,

3.2 *LEMMA.* If rank $F = m \geq 2$ and if $G = F/R$ is finite then the character χ of the relation module $R/R' \otimes_{\mathbb{Z}} Q$ is given by the formula:

$$\chi(fR) = \begin{cases} 1 & \text{if } f \notin R \\ 1+(m-1)|G| & \text{if } f \in R. \end{cases}$$

Proof. If $f \in R$, then $rR'\cdot f = r^fR' = rR'$ and it follows that $\operatorname{trace}(T(fR)) = \operatorname{rank} R = 1+(m-1)|G|$. If $f \notin R$ then we consider the group $S = \langle R,f\rangle$ so that S is free with S/R finite cyclic. By (15) it follows that $\operatorname{trace}(T(fR)) = 1$. □

3.3 *THEOREM.* (Gaschütz 1954). If $G = F/R$ is finite with rank $F = m \geq 2$ then $R/R' \otimes_{\mathbb{Z}} Q \cong Q \oplus \underbrace{QG \oplus \ldots \oplus QG}_{\leftarrow m-1 \rightarrow}$, where QG is the group algebra of G over Q.

Proof. Let χ denote the character of $Q \oplus \underbrace{QG \oplus \ldots \oplus QG}_{m-1}$ as a right G-module. If $g \neq 1$ then the matrix $T(g)$ of the linear transformation induced by the action of g is given by

$$T(g) = I \oplus \underbrace{E \oplus \ldots \oplus E}_{m-1}$$

where E is $|G| \times |G|$ permutation matrix and I is 1×1 identity matrix. Thus $\operatorname{trace}(T(g)) = 1$. If $g = 1$ then $T(g)$ is the $n \times n$ identity matrix where $n = 1+(m-1)|G|$. Thus

$$\chi(g) = \begin{cases} 1 & \text{if } g \neq 1 \\ 1+(m-1)|G| & \text{if } g = 1. \end{cases} \tag{16}$$

Since, by Lemma 3.2, χ is also the character of the relation module $R/R' \otimes_{\mathbb{Z}} Q$, the result follows because two QG-modules are equivalent if they have the same character. □

3.4 *COROLLARY*. If $G = F/R$ is finite with rank $F \neq 1$ then R/R' is a faithful $\mathbb{Z}G$-module, i.e., if $z \in \mathbb{Z}G$ with $rR' \cdot z = 0$ for all $r \in R$ then $z = 0$.

Proof. *Case I* ($\operatorname{rank}(F) = n \geq 2$)

By Theorem 3.3 it follows that if $rR' \cdot z = 0$ for all $r \in R$ then $uz = 0$ in QG for all $u \in QG$ and consequently $z = 0$.

Case II (rank (F) is infinite).

In this case we can choose a set of free generators of F with one free generator x in R. Then $xR' \cdot z = 0$ implies $\alpha_{12}(x)z = 0$ and, in turn, $\lambda_x z = 0$; $z = 0$. □

Corollary 3.4 holds more generally as is shown by the following important result.

3.5 *THEOREM*. (Passi 1975). If R is a normal subgroup of a non-cyclic free group F then R/R' is a faithful $\mathbb{Z}(F/R)$-module.

Proof. We use the right-module homomorphism $R/R' \to \sum_x \lambda_x \mathbb{Z}(F/R)$ given by the Magnus embedding of F/R' via $rR' \to \sum_x \lambda_x \theta \partial_x'(r)$ (see Corollary I.1.13, Remark I.2.3). Let

$$A = \{z \in \mathbb{Z}(F/R) \mid R/R' \cdot z = 0\}.$$

Then for all $u, v \in \mathbb{Z}(F/R)$, $z \in A$,

$$\begin{aligned} R/R' \cdot uzv &= (R/R' \cdot u) \cdot zv \\ &\leq R/R' \cdot zv \\ &= (R/R' \cdot z) \cdot v \\ &= 0, \end{aligned}$$

and it follows that A is a two-sided ideal of $\mathbb{Z}(F/R)$. Let

$$B = \{\theta \partial_x'(r) \mid x \in X,\ r \in R\}.$$

We have,

$$z \in A \text{ if and only if } \theta \partial_x'(r) \cdot z = 0$$

for all $r \in R$, $x \in X$.

It follows that A is an annihilator ideal of $\mathbb{Z}G$ $(G = F/R)$ which annihilates the set B. Let

$$\delta^+(G) = \{g \in G \mid g \text{ has finite order and has only finitely many conjugates}\}.$$

It follows by a theorem of Martha Smith (1971) that if A is an annihilator ideal in a group ring $\mathbb{Z}G$ then $A = (A \cap \mathbb{Z}\,\delta^+(G))\mathbb{Z}G$. Applying this to our situation it suffices to prove that $A \cap \mathbb{Z}\,\delta^+(G) = 0$. Indeed, let $z \in A$ with $z = n_1 g_1 R + \ldots + n_t g_t R$, $n_i \in \mathbb{Z}$, $g_i R \in \delta^+(G)$. Then $H/R = \langle g_1 R, \ldots, g_t R \rangle$ is a finite group (see Passman (1977), page 117). Thus setting $S = \langle R, g_1, \ldots, g_t \rangle$ reduces the problem to S/R' with S/R finite. Further, S is not cyclic since R is not cyclic (being a normal subgroup of the non-cyclic free group F). By Corollary 3.4, S/R' is a faithful $\mathbb{Z}(S/R)$-module and it follows that $z = 0$. □

We close this section with a description of $R/R' \otimes_{\mathbb{Z}} Q$ as a Q-vector space. Thus, let $G = F/R$ be a finite group given by its pre-abelian presentation

$$G = \langle x_1, \ldots, x_m;\ r_1, \ldots, r_q \rangle,\ m \geq 2, \tag{17}$$

where $r_i = x_i^{e_i} \xi_i$ for $i = 1, \ldots, m$; $r_j = \xi_j$, $j = m+1, \ldots, q$; $e_m \mid \ldots \mid e_1 > 0$, $\xi_i \in F'$ (see Magnus et al. (1966), page 140). Then as a Q-vector space,

$$V = R/R' \otimes_{\mathbb{Z}} Q$$

is spanned by all $r_i R' \otimes 1$ $(1 \leq i \leq q)$ where clearly

$$\{r_1R' \otimes 1, \ldots, r_mR' \otimes 1\}$$

is a linearly independent set. Without loss of generality, we may assume that

$$\{r_iR' \otimes 1,\ i = 1, \ldots, q\},\ q \geq m,$$

is a basis of V, where by the Schreier formula

$$q = 1 + (m-1)|G|.$$

For each $i = 1, \ldots, m$, let

$$r_i^*R' = \prod_{g \in G} r_iR' \cdot g.$$

Then

$$r_i^* = r_i^{|G|}\eta_i$$

with $\eta_i \in R \cap F'$. Therefore,

$$r_i^*R' \otimes 1 = |G|(r_iR' \otimes 1) + \eta_iR' \otimes 1.$$

Since $\eta_i \in R \cap F'$, $\eta_iR' \otimes 1$ lies in the subspace

$$W = \{r_jR' \otimes 1;\ j = m+1, \ldots, q\}.$$

We thus have, in addition to Theorem 3.3,

3.6 *THEOREM*. Let $G = F/R$ be a finite group as given by (17). Then

$$R/R' \otimes_{\mathbb{Z}} Q = V = V_1 \oplus \ldots \oplus V_m \oplus W,$$

where V_i, $i = 1, \ldots, m$, is the 1-dimensional subspace spanned by $r_i^*R' \otimes 1$ and W is the $(q-m)$-dimensional subspace spanned by $\{r_jR' \otimes 1;\ j = m+1, \ldots, q\}$. Moreover, G acts trivially on each V_i and W is G-invariant. □

4. RESIDUAL NILPOTENCE OF F/R'

A group $G = F/R$ is said to be residually nilpotent if $\bigcap_m \gamma_m(G) = \{1\}$. A quotient ring $\mathbb{Z}F/\mathfrak{x}$ is said to be residually nilpotent if $\bigcap_m (\mathfrak{x}+\mathfrak{f}^m) = \mathfrak{x}$. With $\mathfrak{r} = \mathbb{Z}F(R-1)$, it is clear that $\mathbb{Z}F/\mathfrak{r}$ is residually nilpotent if and only if $\Delta^{\omega}(F/R) = \bigcap_m \Delta^m(F/R) = 0$, where $\Delta(F/R)$ is the augmentation ideal of $\mathbb{Z}(F/R)$.

4.1 *PROPOSITION*. $\Delta^\omega(F/R) = 0$ implies F/R residually nilpotent.

Proof. Let $\Delta^\omega(F/R) = 0$ and let $fR \in \bigcap_m \gamma_m(F/R)$. Then $f-1 \in \bigcap_m (\underline{\underline{r}}+\underline{\underline{f}}^m) = \underline{\underline{r}}$, by hypothesis. Thus $f \in R$ and $\bigcap_m \gamma_m(F/R) = \{1R\}$. □

4.2 *THEOREM*. (Passi 1975, Lichtman 1977). F/R' is residually nilpotent if and only if $\Delta^\omega(F/R) = 0$.

Proof. Let $\Delta^\omega(F/R) = 0$ and consider the Magnus embedding

$$\psi\colon F/R' \to \begin{bmatrix} F/R & \sum_x \mathbb{Z}(F/R)\lambda_x \\ 0 & 1R \end{bmatrix}$$

as given by I.1.10.

If $w \in \gamma_m(F/R')$ then it is easy to see by a simple induction on m that

$$\psi(w) \in \begin{bmatrix} \gamma_m(F/R) & \sum_x \Delta^{m-1}(F/R)\lambda_x \\ 0 & 1R \end{bmatrix}.$$

Thus $w \in \bigcap_m \gamma_m(F/R')$ implies

$$\psi(w) \in \begin{bmatrix} \bigcap_m \gamma_m(F/R) & \sum_x \Delta^\omega(F/R)\lambda_x \\ 0 & 1R \end{bmatrix}$$

$$= \begin{bmatrix} 1R & 0 \\ 0 & 1R \end{bmatrix}$$

by Proposition 4.1 and by hypothesis. It follows that $w \in R'$ and consequently F/R' is residually nilpotent. Thus $\Delta^\omega(F/R) = 0$ implies F/R' residually nilpotent.

Conversely, let F/R' be residually nilpotent. For $r \in R/R'$ and $u \in \Delta^m(F/R)$, $r\cdot u \in \gamma_{m+1}(F/R')$. Thus if $z \in \bigcap_m \Delta^m(F/R)$, then $R/R'\cdot z \le \bigcap_m \gamma_m(F/R') = \{1R'\}$, by hypothesis. Thus by Theorem 3.5 it follows that $z = 0$ and, in turn, $\Delta^\omega(F/R) = 0$. □

Residual nilpotence of F/R' has been characterized further by Lichtman (1977). We need some definitions.

If P is a group property, then a group G is said to be residually P if for every $g \in G\setminus\{1\}$, there exists a normal subgroup N_g of G such that G/N_g has the property P and $g \notin N_g$.

If $\underline{\underline{C}}$ is a class of groups, then a group G is said to be discriminated by $\underline{\underline{C}}$ if for every finite set $\{g_1,\ldots,g_n\}$ of distinct elements of G, there exists a group $H \in \underline{\underline{C}}$ and a homomorphism $\alpha: G \to H$ such that $\alpha(g_1),\ldots,\alpha(g_n)$ are distinct elements of H.

We conclude by stating the following characterization of $\Delta^{\omega}(G) = 0$.

4.3 *THEOREM*. (Lichtman 1977). Let G be a group. Then $\Delta^{\omega}(G) = 0$ if and only if either (i) G is residually torsion-free nilpotent; or (ii) G is discriminated by the class of nilpotent groups of finite prime power exponents. □

[For proof see also Passi (1979), page 92. See also Hartley (1970) and (1984).]

Chapter III

Fox Subgroups of Free Groups

Let F be a free group and R a normal subgroup of F. Recall from Chapter I that, for each $n \geq 1$, the n-th Fox subgroup of F relative to R is defined by

$$F(n,R) = F \cap (1+\mathfrak{r}\mathfrak{f}^n)$$

where $\mathfrak{f} = \mathbb{Z}F(F-1)$ and $\mathfrak{r} = \mathbb{Z}F(R-1)$ are ideals of $\mathbb{Z}F$. Also recall that $F(1,R) = R'$ (Corollary I.1.9). In this chapter we give a complete identification of $F(n,R)$, $n \geq 2$, yielding a solution of the Fox problem as stated in I.2.6. For finite F/R, we also determine the structure of the Fox modules $(F(n,R)/F(n+1,R)) \otimes_{\mathbb{Z}} \mathbb{Q}$ for $n \geq 1$.

1. A FAITHFUL MATRIX REPRESENTATION OF $F/F(n,R)$

Let F be free on a set X of cardinality exceeding one. For each $n \geq 1$, let

$$\Lambda_n = \{\lambda^{(x)}_{i,i+1};\ 1 \leq i \leq n,\ x \in X\},$$

be a set of independent and commuting indeterminates and let $\Omega_{n+1} = \mathbb{Z}(F/R)[\Lambda_n]$ be the ring of polynomials in $\lambda^{(x)}_{i,i+1}$'s over the integral group ring $\mathbb{Z}(F/R)$. We set $\Omega_1 = \mathbb{Z}(F/R)$ and denote by M_{n+1}, $n \geq 0$, the ring of all $n+1 \times n+1$ upper-triangular matrices over Ω_{n+1}. For each $x \in X$, the matrix

$$\psi_{n+1}(x) = \begin{bmatrix} xR & \lambda^{(x)}_{12} & 0 & \cdots & 0 \\ 0 & 1R & \lambda^{(x)}_{23} & \cdots & 0 \\ 0 & 0 & 1R & \cdots & 0 \\ \vdots & & & & \\ 0 & 0 & 0 & \cdots & 1R \end{bmatrix}$$

is an invertible element of M_{n+1} and the map

$$\psi_{n+1}: X \to M_{n+1}$$

given by $x \to \psi_{n+1}(x)$ defines a homomorphism of F into $U(M_{n+1})$, the group of units of M_{n+1}. We extend ψ_{n+1} by linearity to a ring homomorphism

$$\psi^*_{n+1}: \mathbb{Z}F \to M_{n+1}.$$

For $u \in \mathbb{Z}F$, we denote by $\alpha_{ij}(u)$, $1 \le i \le j \le n+1$; the ij-entry of $\psi^*_{n+1}(u)$. [Note that, for $m \ge n$, ij-entry of $\psi^*_{n+1}(u)$ = ij-entry of $\psi^*_{m+1}(u)$, i.e., $\alpha_{ij}(u)$ is independent of n.] Observe that for $i \ge 2$, $\alpha_{ii}: \mathbb{Z}F \to \mathbb{Z}$ is the augmentation map and $\alpha_{11}: \mathbb{Z}F \to \mathbb{Z}(F/R)$ is the linear extension of natural projection: $F \to F/R$. We prove,

1.1 *THEOREM*. (Gupta & Passi 1976). For all $n \ge 0$, $\ker \psi^*_{n+1} = \underline{\underline{r}}\underline{\underline{f}}^n$.

Proof. We proceed by induction on n. For $n = 0$, $M_1 = \mathbb{Z}(F/R)$ and $\psi^*_1: \mathbb{Z}F \to \mathbb{Z}(F/R)$ is the natural projection. Hence, $\ker \psi^*_1 = \underline{\underline{r}} = \underline{\underline{r}}\underline{\underline{f}}^0$. Let $n \ge 1$ and assume the result for $n-1$. We first prove that $\underline{\underline{r}}\underline{\underline{f}}^n \le \ker \psi^*_{n+1}$. Let $u = v(w-1)$ with $v \in \underline{\underline{r}}\underline{\underline{f}}^{n-1}$, $w \in F$. Since $\underline{\underline{r}}\underline{\underline{f}}^n$ is spanned by elements of the form $u = v(w-1)$, it suffices to prove that $\alpha_{ij}(u) = 0$ for $1 \le i \le j \le n+1$. Indeed, by the induction hypothesis $\psi^*_n(v) = 0 = \psi^*_n(u)$, so $\alpha_{ij}(u) = 0$ for $1 \le i \le j \le n$. Further, for $1 \le i \le n+1$,

$$\begin{aligned}\alpha_{i,n+1}(u) &= \sum_{k=i}^{n+1} \alpha_{i,k}(v)\alpha_{k,n+1}(w-1)\\ &= \alpha_{i,n+1}(v)\alpha_{n+1,n+1}(w-1)\\ &= 0,\end{aligned}$$

since $\psi^*_n(v) = 0$ and $\alpha_{n+1,n+1}(w-1) = 0$. This proves $\underline{\underline{r}}\underline{\underline{f}}^n \le \ker \psi^*_{n+1}$.

For the reverse inclusion, let $z \in \ker \psi^*_{n+1} \le \ker \psi^*_1 \le \underline{\underline{f}}$. Then $z = \sum_x \mu(x)(x-1)$ for some finitely many non-zero elements $\mu(x)$ in $\mathbb{Z}F$. Now, for $1 \le k \le \ell \le n$, we have

$$0 = \alpha_{k,\ell+1}(z) = \sum_x \sum_{j=k}^{\ell+1} \alpha_{k,j}(\mu(x))\alpha_{j,\ell+1}(x-1)$$

$$= \sum_x \alpha_{k,\ell}(\mu(x))\alpha_{\ell,\ell+1}(x-1),$$

since $\alpha_{j,j'}(x-1) = 0$ for $j'-j \geq 2$. Since $\alpha_{\ell,\ell+1}(x-1) = \lambda^{(x)}_{\ell,\ell+1}$, it follows that $\alpha_{k,\ell}(\mu(x)) = 0$ for each $x \in X$ and each $1 \leq k \leq \ell \leq n$. Thus, by the induction hypothesis, $\mu(x) \in \underline{\underline{\mathfrak{r}\mathfrak{f}}}^{n-1}$ and consequently $z \in \underline{\underline{\mathfrak{r}\mathfrak{f}}}^{n}$. □

Since $\ker \psi_{n+1} = F \cap (1+\ker \psi^*_{n+1})$, we obtain

1.2 *COROLLARY*. (Gupta & Gupta 1974). $F/F(n,R) \cong \langle\psi_{n+1}(x);\ x \in X\rangle$ for all n. □

1.3 *COROLLARY*. $F/F(n,R)$ is always torsion-free for $n \geq 1$. [The proof follows from II.1.1 together with the fact that $R/F(n,R)$ is torsion-free.] □

2. IDENTIFICATION OF $F(2,R)$

Let F be free on X and R, as a subgroup of F, be free on $\underline{Y} \subseteq F$. Set $Y = Y_1 \cup Y_2$, where $Y_1 \cap F' = \phi$ and $Y_2 \subseteq F'$. If $w \in \gamma_3(R)$ then $w-1 \in \underline{\mathfrak{r}}^3 \leq \underline{\underline{\mathfrak{r}\mathfrak{f}}}^2$ and if $w \in [R \cap F', R \cap F']$ then $w-1 \in (\underline{\mathfrak{r}} \cap \underline{\mathfrak{f}}^2)(\underline{\mathfrak{r}} \cap \underline{\mathfrak{f}}^2) \leq \underline{\underline{\mathfrak{r}\mathfrak{f}}}^2$. It follows that if $G(2,R) = [R \cap F', R \cap F']\gamma_3(R)$ then $G(2,R) \leq F(2,R)$.

2.1 *THEOREM*. (Enright 1968, Hurley 1973). $F(2,R) = G(2,R)$.

Proof. Let $w \in F(2,R)$. The proof consists in showing that $w \in G(2,R) = [R \cap F', R \cap F']\gamma_3(R)$. Since $F(2,R) \leq F(1,R) \cap (F \cap (1+\underline{\mathfrak{f}}^3)) = R' \cap \gamma_3(F)$ (Corollary I.1.9, Theorem I.3.7), it follows that $w \in \gamma_3(F)$ and so $w \equiv \prod_{i,j} [y_{1i},y_{2j}]^{a_{ij}} \pmod{G(2,R)}$, where $a_{ij} \in \mathbb{Z}$, $y_{1i} \in Y_1$, $y_{2j} \in Y_2$. Thus, modulo $G(2,R)$, we may write

$$w \equiv \prod_j [z_{1j},y_{2j}],$$

where $z_{1j} \in \langle Y_1\rangle \leq R$ and $z_{1j} \notin R \cap F'$. This yields, modulo $\underline{\underline{\mathfrak{r}\mathfrak{f}}}^2$,

$$w-1 \equiv \sum_j -(y_{2j}-1)(z_{1j}-1) \equiv 0.$$

Since Y_2 is a part of a free basis for R and $\underline{\mathfrak{r}}$ is a free right $\mathbb{Z}F$-module on $\{y-1;\ y \in Y\}$ (Proposition I.1.12), it follows that $z_{1j}-1 \in \underline{\mathfrak{f}}^2$ for all j and consequently

$z_{1j} \in F' \cap R$, contrary to our choice of z_{1j}. Thus $w \equiv 1 \pmod{G(2,R)}$. □

3. IDENTIFICATION OF $F(n,R)$ (F/R FINITE)

Let $G = F/R$ be a finite group given by its pre-abelian presentation:

$$G = \langle x_1,\ldots,x_m;\ x_1^{e_1}\xi_1,\ldots,x_m^{e_m}\xi_m,\xi_{m+1},\ldots\rangle\ , \qquad (1)$$

where $e_m|\ldots|e_1 > 0$, $\xi_1 \in F'$ (see for instance Magnus et al. (1966), page 140).

For $S \trianglelefteq R$ with R/S nilpotent, define

$$\sqrt{S} = \{r \in R \mid r^{n(r)} \in S \text{ for some } n(r) \in \mathbb{Z}\setminus\{0\}\}$$

to be the isolator of S in R. Then $R/\sqrt{S}$ is a torsion-free nilpotent group. For all $t \geq 1$ define

$$R_t = R \cap \gamma_t(F).$$

3.1 *LEMMA*. $R_t \leq \sqrt{\gamma_t(R)\,R_{t+k}}$ for all $t,k \geq 1$.

Proof. Let $u = [x_{i(1)},\ldots,x_{i(t)}]$ and consider the element $v = [x_{i(1)}^{e_{i(1)}}\xi_{i(1)},\ldots,x_{i(t)}^{e_{i(t)}}\xi_{i(t)}]$ which by (1) lies in $\gamma_t(R)$. Then, with $e = e_{i(1)}\cdots e_{i(t)} \neq 0$, $v \equiv u^e \pmod{\gamma_{t+1}(F)}$; or equivalently $u \in \sqrt{\gamma_t(R)\gamma_{t+1}(F)}$. It follows that $\gamma_t(F) \leq \sqrt{\gamma_t(R)\gamma_{t+1}(F)}$ and consequently,

$$R_t = \gamma_t(F) \cap R \leq \sqrt{\gamma_t(R)\gamma_{t+1}(F)} \cap R \leq \sqrt{\gamma_t(R)R_{t+1}}\ .$$

Thus, by iteration,

$$R_t \leq \sqrt{\gamma_t(R)R_{t+1}} \leq \sqrt{\gamma_t(R)R_{t+2}} \leq \ldots\ .$$ □

3.2 *THEOREM*. (Gupta 1977). If F/R is finite then $F(n,R) = \sqrt{[R_n,R_n]\gamma_{n+1}(R)}$ for all $n \geq 1$.

Proof. Put $E(n,R) = [R_n,R_n]\gamma_{n+1}(R)$. Then, since $[R_n,R_n]-1 \leq (\mathfrak{r} \cap \mathfrak{f}^n)(\mathfrak{r} \cap \mathfrak{f}^n) \leq \mathfrak{r}\mathfrak{f}^n$ and $\gamma_{n+1}(R) \leq \mathfrak{r}^{n+1} \leq \mathfrak{r}\mathfrak{f}^n$, it follows that $E(n,R) \leq F(n,R)$ and since $F/F(n,R)$ is torsion-free (Corollary 1.3), $\sqrt{E(n,R)} \leq F(n,R)$. It remains to prove by induction on $n \geq 1$ that $F(n,R) \leq \sqrt{E(n,R)}$. When $n = 1$,

$F(1,R) = R'$ (Corollary I.1.9) and, since $E(1,R) = R' = \sqrt{R'}$ (Theorem II.1.1), $F(1,R) \leq \sqrt{E(1,R)}$. Let $n > 1$ and assume the result for $n-1$. Then

$$F(n,R) \leq F(n-1,R) \leq \sqrt{[R_{n-1},R_{n-1}]\gamma_n(R)}\ .$$

By Lemma 3.1, $R_{n-1} \leq \sqrt{\gamma_{n-1}(R)R_n}$. Since $R_{n-1}/[R_n,R_n]\gamma_n(R)$ is nilpotent and $[\gamma_{n-1}(R)R_n,\gamma_{n-1}(R)R_n] \leq [R_n,R_n]\gamma_n(R)$, it follows that

$$[R_{n-1},R_{n-1}] \leq \sqrt{[R_n,R_n]\gamma_n(R)}\ .$$

Thus

$$F(n,R) \leq \sqrt{\gamma_n(R)E(n,R)}\ . \tag{2}$$

Let $w \in F(n,R)$ with $w \notin \sqrt{E(n,R)}$. Then, by (2), some power of w lies in $\gamma_n(R) \cap F(n,R) \leq \gamma_n(R) \cap \gamma_{n+1}(F)$, since by Theorem I.3.7,

$$F(n,R) \leq F \cap (1+\underline{\mathfrak{f}}^{n+1}) = \gamma_{n+1}(F).$$

Thus, without loss of generality, we may assume that w itself lies in $\gamma_n(R) \cap \gamma_{n+1}(F) \leq [R_2,\underbrace{R,\ldots,R}_{n-1}]$. Let $z = [z_1,z_2,\ldots,z_n]$ be a factor of w with $z_1 \in R_2$, $z_i \in R$, $i = 2,\ldots,n$. If, for some $i = 2,\ldots,n$, $z_i \in R \cap F'$, then, by Lemma 3.1, some power of z will lie in $[R_n,R_n]\gamma_{n+1}(R)$. Thus, if R is free on $Y = Y_1 \cup Y_2$, $Y_1 \cap F' = \phi$ and $Y_2 \subseteq F'$, then we may assume that w is a product of the form

$$w = \prod_{\underset{\sim}{i}} [z_{\underset{\sim}{i}},y_{1i(1)},\ldots,y_{1i(n-1)}]$$

where $y_{1j} \in Y_1$, $z_{\underset{\sim}{i}} \in R_2\backslash R'$ and the product is taken over <u>distinct</u> $(n-1)$-tuples $\underset{\sim}{i} = (i(1),\ldots,i(n-1))$.

With $y_{ij} = x_{ij}^{e_{ij}}\xi_{ij}$, $x_{ij} \in \{x_1,\ldots,x_m\}$, expansion of $w-1$ yields

$$\sum_{\underset{\sim}{i}} (z_{\underset{\sim}{i}}-1)(y_{1i(1)}-1)\ldots(y_{1i(n-1)}-1) \equiv 0 \pmod{\underline{\underline{\mathfrak{r}\mathfrak{f}}}^n}$$

which, in turn, yields

$$\sum_{\underset{\sim}{i}} e_{\underset{\sim}{i}}(z_{\underset{\sim}{i}}-1)(x_{1i(1)}-1)\ldots(x_{1i(n-1)}-1) \equiv 0 \pmod{\underline{\underline{\mathfrak{r}\mathfrak{f}}}^n}$$

where

$$e_{\underset{\sim}{i}} = e_{1i(1)} \cdots e_{1i(n-1)} \neq 0.$$

Since $\underline{\underline{\delta}}^{n-1}$ is a left $\mathbb{Z}F$-module on distinct products of the form $(x_{1i(1)}-1)\cdots(x_{1i(n-1)}-1)$ (Remark I.1.11), it follows that $z_{\underset{\sim}{i}}-1 \in \underline{\underline{r\delta}}$ and consequently $z_{\underset{\sim}{i}} \in R'$ (Corollary I.1.9), contrary to the choice of $z_{\underset{\sim}{i}}$'s. Thus $w \in \sqrt{E(n,R)}$ as was to be proved. □

4. A FILTRATION OF THE LOWER CENTRAL SERIES OF F

For the purpose of studying the general Fox subgroup $F(n,R)$ it is essential to study the structure of the $Q(F/R)$-module $(\gamma_k(R)/\gamma_{k+1}(R) \otimes_{\mathbb{Z}} Q$ in terms of the lower central factors of F. The basis theorem of P. Hall (see Magnus et al. (1966), or Theorem I.3.8) allows us to write an arbitrary element of a free nilpotent group as a product of certain "basic" commutators in a canonical fashion.

Let F be a finitely generated free group, $d(i)$, $i \geq 1$, the rank of $\gamma_i(F)/\gamma_{i+1}(F)$ and

$$B_i = \{b_{i,1} < \cdots < b_{i,d(i)}\}, \tag{3}$$

be an ordered set of basic commutators of weight i. Between $\gamma_i(F)$ and $\gamma_{i+1}(F)$ we introduce normal subgroups $\gamma_{i,j}(F)$, $j = 1,\ldots,d(i)$, of F as follows:

$$\gamma_i(F) = \gamma_{i,1}(F) \rhd \gamma_{i,2}(F) \rhd \cdots \rhd \gamma_{i,d(i)}(F) \rhd \gamma_{i+1}(F) \tag{4}$$

where $\gamma_{i,j}(F)/\gamma_{i,j+1}(F)$ is the infinite cyclic group generated by the coset $b_{i,j}\gamma_{i,j+1}(F)$ (here $\gamma_{i,d(i)+1}(F) = \gamma_{i+1}(F) = \gamma_{i+1,1}(F)$). For convenience, we write

$$\gamma_{i,j}^{+}(F) = \gamma_{i,j+1}(F).$$

Since $\bigcap_n \gamma_n(F) = \{1\}$, for any $w \in F$, $w \neq 1$, there exists a unique pair (i,j) of positive integers such that $w \in \gamma_{i,j}(F)$ and $w \notin \gamma_{i,j}^{+}(F)$. We order $\{(i,j),\ i \leq j \leq d(i)\}$ lexicographically. For $u \in \gamma_{i,j}(F)\setminus\gamma_{i,j}^{+}(F)$, $v \in \gamma_{i',j'}(F)\setminus\gamma_{i',j'}^{+}(F)$, we define

$$u <^* v \quad \text{if} \quad (i,j) < (i',j')$$

$$u \overline{\underline{*}}\ v \quad \text{if} \quad (i,j) = (i',j'). \tag{5}$$

With respect to the $*$-ordering, F is a totally-ordered set.

Let R be a proper normal subgroup of F. For each $k \geq 1$, the quotient

$$\overline{\gamma_k(R)} = \gamma_k(R)/\gamma_{k+1}(R)$$

is a right F/R-module via conjugation in F. The right $Q(F/R)$-module

$$M_k = \overline{\gamma_k(R)} \otimes_{\mathbb{Z}} Q \tag{6}$$

has a basis $\bar{Y}_k$ consisting of elements of the form $y_{k,(i,j)}\gamma_{k+1}(R)$, where

$$y_{k,(i,j)} \in \gamma_k(R) \cap \gamma_{i,j}(F) \backslash \gamma_k(R) \cap \gamma^+_{i,j}(F).$$

Set

$$Y_k = \{y_{k,(i,j)} \mid y_{k,(i,j)}\gamma_{k+1}(R) \in \bar{Y}_k\}. \tag{7}$$

[Note that $i \geq k$.]

For each $n \geq 1$, let

$$Y_k^{(n)} = \{y_{k,(i,j)} \in Y_k \mid k \leq i \leq n\}. \tag{8}$$

Then $Y_k^{(n)}$ is a (possibly empty) finite set.

Finally, put

$$U^{(n)} = \bigcup_{k=1}^{n} Y_k^{(n)}. \tag{9}$$

If, for some $\ell > k$,

$$y_{k,(i,j)} \mathrel{\bar{\ast}} y_{\ell,(i,j)}$$

then since $\gamma_{i,j}(F)/\gamma^+_{i,j}(F)$ is cyclic, we have for some $a,b \in \mathbb{Z}\backslash\{0\}$,

$$y^a_{k,(i,j)}y^b_{\ell,(i,j)} \in \gamma^+_{i,j}(F).$$

Thus, replacing $y_{k,(i,j)}\gamma_{k+1}(R)$ by $y^a_{k,(i,j)}y^b_{\ell,(i,j)}\gamma_{k+1}(R)$, if necessary, we may assume that no two elements of $U^{(n)}$ are related by $\bar{\ast}$.

Let $\bar{U}^{(n)} = \{\bar{y}_{k,(i,j)} = y_{k,(i,j)}\gamma_{k+1}(R) \mid y_{k,(i,j)} \in U^{(n)}\}$. Then we have proved the following,

4.1 LEMMA. (Gupta 1984).

(i) $\bar{U}^{(n)}$ is a basis of the right $\mathbb{Z}(F/R)$-module

$$\sum_{k=1}^{n} \frac{\gamma_k(R)\ \gamma_{n+1}(F)}{\gamma_{k+1}(R)\ \gamma_{n+1}(F)} \otimes_{\mathbb{Z}} Q .$$

(ii) $\bar{Y}_1^{(n)}$ is a basis of the right $\mathbb{Z}(F/R)$-module

$$(R\ \gamma_{n+1}(F)/R'\ \gamma_{n+1}(F)) \otimes_{\mathbb{Z}} Q . \qquad \square$$

Since the basic products of the form

$$(b_{i_1,j_1}-1) \ldots (b_{i_k,j_k}-1)$$

with $(i_1,j_1) \le \ldots \le (i_k,j_k)$ and $i_1 + \ldots + j_k = n$ are linearly independent modulo $\mathfrak{f}^{n+1}$, we have the following immediate corollary of Lemma 4.1.

4.2 LEMMA. The elements of the form

$$(y_{m_1},(i_1,j_1)^{-1}) \ldots (y_{m_\ell},(i_\ell,j_\ell)^{-1})$$

with $m_1 + \ldots + m_\ell = k$, $(i_1,j_1) \le \ldots \le (i_\ell,j_\ell)$ and $i_1 + \ldots + i_\ell = n$, form a part of the basis for $(\mathfrak{r}^k \cap \mathfrak{f}^n + \mathfrak{r}^{k+1} + \mathfrak{f}^{n+1}/\mathfrak{r}^{k+1} + \mathfrak{f}^{n+1}) \otimes_{\mathbb{Z}} Q$. $\square$

5. A SOLUTION OF THE FOX PROBLEM

For $t \ge 1$, let $R_t = R \cap \gamma_t(F)$. Then $R_t - 1 \le \mathfrak{r} \cap \mathfrak{f}^t$ and so $[R_{t_1},R_{t_2}]-1 \le (\mathfrak{r} \cap \mathfrak{f}^{t_1})(\mathfrak{r} \cap \mathfrak{f}^{t_2}) + (\mathfrak{r} \cap \mathfrak{f}^{t_2})(\mathfrak{r} \cap \mathfrak{f}^{t_1}) \le \mathfrak{r}\mathfrak{f}^t$, $t = \min.\{t_1,t_2\}$. More generally, for $n \ge 1$, $2 \le m \le n+1$, let

$$\underset{\sim}{t}(m) = (t_1,\ldots,t_m) \tag{10}$$

be an m-tuple of positive integers satisfying

$$t_1 + \ldots + \hat{t}_i + \ldots + t_m \ge n,\ i = 1,\ldots,m.$$

($\hat{t}_i$ indicates t_i missing.) Then it is readily verified that

$$[R_{t_1},\ldots,R_{t_m}] \le F \cap (1+\mathfrak{r}\mathfrak{f}^n) = F(n,R).$$

In particular, if

$$G(n,R) = \sum_{m=2}^{n+1} \prod_{\underset{\sim}{t}(m)} [R_{t_1},\ldots,R_{t_m}] \tag{11}$$

where the product is taken over all $\underset{\sim}{t}(m)$ satisfying (10), then $G(n,R) \leq F(n,R)$.

With $R = F'$, it was proved by Gupta & Gupta (1974) that $F(n,F') = G(n,F')$ for all n. For the general case we note by Corollary 1.3 that $F/F(n,R)$ is always torsion-free. Thus we must replace $G(n,R)$ by its isolator $\sqrt{G(n,R)}$ in R and ask: Is $F(n,R) = \sqrt{G(n,R)}$? In an attempt to answer this question it appeared that the solution hinged on another related problem. Since $[R_i,R_j] \leq R' \cap \gamma_{i+j}(F)$, it follows that if

$$H(n,R) = \prod_{i+j=n} [R_i,R_j] ,$$

then $H(n,R) \leq R' \cap \gamma_n(F)$. Again replacing $H(n,R)$ by $\sqrt{H(n,R)}$ we ask: Is $R' \cap \gamma_n(F) = \sqrt{H(n,R)}$? It was announced in Gupta (1981) that if, for some $n \geq 2$, $R' \cap \gamma_n(F) = \sqrt{H(n,R)}$ then $F(n+1,R) = \sqrt{G(n+1,R)}$. The details appeared in Gupta (1984) (submitted 1981), where in addition the equality $R' \cap \gamma_n(F) = \sqrt{H(n,R)}$ was proved for small values of n (e.g., $n \leq 6$). Recently, Hurley (1985) reports that the above equality holds for all n and hence $F(n,R) = \sqrt{G(n,R)}$ for all n.

Yunus (1984) offers an independent solution of the Fox problem without using $H(n,R)$ (collecting words modulo $\gamma_{k+1}(R) \cap F(n,R)$ as opposed to $\gamma_{k+1}(R) \cap \gamma_{n+1}(F)$ in Gupta (1984) which necessarily invoked $H(n,R)$). Once the assertion of the main step (Lemma 4.3 below) is formulated without $H(n,R)$, it turns out that the proof given in Gupta (1984) is actually a proof of the new assertion. In what follows we shall reformulate our assertion as in Yunus (1984) and reproduce its proof from Gupta (1984).

For each $1 \leq m \leq n$, we define $P_m(n,R) \trianglelefteq F$ by

$$P_m(n,R) = \prod_{\underset{\sim}{t}(m)} [R^*_{t_1},\ldots,R^*_{t_m}] \tag{12}$$

where $R^*_t = \sqrt{R_tR'}$, $R_t = R \cap \gamma_t(F)$ and the product is taken over all m-tuples $\underset{\sim}{t}(m)$ satisfying (10).

The proof of $F(n,R) = \sqrt{G(n,R)}$ will be seen to follow from the following fundamental lemma whose formulation is due to

Yunus (1984).

5.1 *LEMMA*. (Yunus 1984). Let R be a normal subgroup of a finitely generated free group F. Then for all $k \geq 2$,

$$\gamma_k(R) \cap F(n,R) \leq \sqrt{P_k(n,R)\ \gamma_{k+1}(R)} .$$

Proof. (Gupta 1984). By induction on $n \geq 1$. When $n = 1$, $F(1,R) = R'$ (Corollary I.1.9) and $P_k(1,R) = \gamma_k(R)$, so there is nothing to prove. Assume the result for some $n \geq 1$. Let $w \in \gamma_k(R) \cap F(n+1,R)$. Since $F(n+1,R) \leq F(n,R)$, it follows by the induction hypothesis that $w \in \sqrt{P_k(n,R)\ \gamma_{k+1}(R)}$. Thus replacing w by a suitable power, if necessary, we may assume that

$$w \in P_k(n,R)\ \gamma_{k+1}(R) \cap F(n+1,R).$$

Since $w \in F(n+1,R)$, it clearly suffices to prove that $w \in \sqrt{P_k(n+1,R)\ \gamma_{k+1}(R)}$. Modulo $P_k(n+1,R)\ \gamma_{k+1}(R)$, w is a product of commutators of the form

$$u = [u_1,\ldots,u_k], \tag{13}$$

satisfying

(a) $u_i \in R_{t_i},\ u_i \notin R'$;

(b) $t_1 + \ldots + \hat{t}_i + \ldots + t_k \geq n$, for all i ;

(c) $t_1 + \ldots + \hat{t}_p + \ldots + t_k = n$, for some $p \leq k$.

Let $u = [u_1,\ldots,u_k]$ be a commutator satisfying (13)(a),(b),(c). Then modulo $P_k(n+1,R)\ \gamma_{k+1}(R)$,

$$\begin{aligned} u &\equiv [u_p,[u_1,\ldots,u_{p-1}],u_{p+1},\ldots,u_k]^{-1} \\ &\equiv \prod_\sigma [u_p,u_{1\sigma},\ldots,u_{(p-1)\sigma},u_{p+1},\ldots,u_k]^{\varepsilon(\sigma)}, \quad \text{(Jacobi)} \end{aligned}$$

where σ is a permutation of $\{1,\ldots,p-1\}$ and $\varepsilon(\sigma) \in \mathbb{Z}$.

Thus, instead of (13) we may assume that w is a product, modulo $P_k(n+1,R)\ \gamma_{k+1}(R)$, of commutators of the form

$$v = [v_0,v_1,\ldots,v_{k-1}], \tag{14}$$

satisfying

(a) $v_i \in R_{t_i}$, $v_i \notin R'$;

(b) $v_i <^* v_0$ or $v_i \,\overline{\underline{*}}\, v_0$ for $i = 1,\dots,k-1$

(given by (5));

(c) $t_1 + \dots + t_{k-1} = n.$

Now, suppose $v = [v_0, v_1, \dots, v_i, v_{i+1}, \dots, v_{k-1}]$ is a factor of w satisfying (14)(a),(b),(c), with $v_{i+1} <^* v_i$, for some i. Then

$$v \equiv [v_0, v_1, \dots, v_{i+1}, v_i, \dots, v_{k-1}]$$

$$\equiv [v_0, v_1, \dots, [v_i, v_{i+1}], \dots, v_{k-1}],$$

where $[v_i, v_{i+1}] \in \gamma_2(R) \cap \gamma_{t_i+t_{i+1}}(F)$. By the repeated application of this procedure, we may rewrite w as a product of commutators of the form

$$v = [v_0, y_{m(1),(i(1),j(1))}, \dots, y_{m(\ell),(i(\ell),j(\ell))}], \tag{15}$$

satisfying

(a) $y_{m,(i,j)} \in U^{(n)}$ (given by (9));

(b) $m(1) + \dots + m(\ell) = k-1$;

(c) $i(1) + \dots + i(\ell) = n$;

(d) $(i(1),j(1)) \le \dots \le (i(\ell),j(\ell))$;

(e) $v_0 \in R \cap \gamma_{t_0}(F)$, $v_0 \notin R'$.

Since $P_k(n+1,R)-1 \le \underline{\underline{r}}\underline{\underline{f}}^{n+1}$ and $\gamma_{k+1}(R)-1 \le \underline{r}^{k+1}$, $w \in \gamma_k(R) \cap F(n+1,R)$ implies that, modulo $\underline{\underline{r}}\underline{\underline{f}}^{n+1} + \underline{r}^{k+1}$, $w-1$ is a sum of Lie elements of the form

$$v^* = ((v_0-1, y_{m(1),(i(1),j(1))}-1, \dots, y_{m(\ell),(i(\ell),j(\ell))}-1))$$

satisfying (15)(a)-(e).

We note that in the expansion of v^*, modulo $\underline{\underline{r}}\underline{\underline{f}}^{n+1}$ there is a unique term

$$(v_0-1)(y_{m(1),(i(1),j(1))}-1) \dots (y_{m(\ell),(i(\ell),j(\ell))}-1) \tag{16}$$

which, in turn, yields the term

$$a(v_0-1)(b_{i(1),j(1)}-1) \dots (b_{i(\ell),j(\ell)}-1),$$

where $y_{k,(i,j)} \equiv b_{i,j}^{a_{ij}}(\lambda_{i,j}^{+}(F))$, $\quad a = \prod_{q=1}^{\ell} a_{i(q),j(q)} \in \mathbb{Z}\setminus\{0\}$.

Let v^* be a summand of $w-1$ as above. Then the segment

$$(y_{m(1),(i(1),j(1))}-1) \ldots (y_{m(\ell),(i(\ell),j(\ell))}-1)$$

of (16) cannot come from any other summand of $w-1$ and since this segment is a part of a basis for $\mathfrak{r}^{k-1} \cap \mathfrak{f}^{n}$ modulo $\mathfrak{r}^{k} + \mathfrak{f}^{n+1}$ (Lemma 4.2), it cannot come from elements of $\mathfrak{r}^{k+1} + \mathfrak{r}\mathfrak{f}^{n+1}$. Since $w-1 \in \mathfrak{r}\mathfrak{f}^{n+1}$, it follows that we must have

$$a(v_0-1)(b_{i(1),j(1)}-1) \ldots (b_{i(\ell),j(\ell)}-1) \in \mathfrak{r}\mathfrak{f}^{n+1}$$

which, in turn, yields

$$(v_0^a-1) \in \mathfrak{r}\mathfrak{f},$$

since $(b_{i(1),j(1)}-1)\ldots(b_{i(\ell),j(\ell)}-1)$ is a basic product of degree n. This yields $v_0^a \in R'$, contrary to (15)(e). Thus $w \in P_k(n+1,R)\ \gamma_{k+1}(R)$ as required. □

We can now deduce the main result of this section.

5.2 *THEOREM*. Let R be a normal subgroup of a free group F. Then $F(n,R) = \sqrt{G(n,R)}$ for all n.

Proof. Since by (11), $G(n,R) \le F(n,R)$ and since $F/F(n,R)$ is torsion-free (Corollary 1.3), it follows that

$$\sqrt{G(n,R)} \le F(n,R)$$

for all n.

For the reverse inequality, since

$$\sqrt{G(n,R)} = \prod_{m=2}^{n+1} \sqrt{P_m(n,R)}$$

by (11), (12) and since

$$P_m(n,R) \le F(n,R) \cap \gamma_m(R),$$

it follows that if $F(n,R) \nleq \sqrt{G(n,R)}$ for some $n \ge 2$, then there is a least integer $k \ge 2$ such that

$$F(n,R) \cap \gamma_k(R) \nleq \sqrt{P_k(n,R)\ \gamma_{k+1}(R)}\ ,$$

contrary to Lemma 5.1. This completes the proof of the main theorem. □

5.3 *REMARKS*. (i) Identification of Fox subgroups appears as Problem 13 in Birman (1975) where the problem is predicted as "a messy problem with an unpleasant answer" (page 115) (see also Lyndon (1984), Problem 20).
(ii) As defined by (11), $G(n,R)$ may contain several redundant factors which can be eliminated by using P. Hall's three-subgroup Lemma ($[A,B,C] \le [B,C,A][C,A,B]$, $A,B,C \trianglelefteq F$). Thus, for example,

$$G(1,R) = [R_1,R_1] = R' ;$$

$$G(2,R) = [R_2,R_2][R_1,R_1,R_1] ;$$

$$G(3,R) = [R_3,R_3][R_2,R_1,R_2][R_1,R_1,R_1,R_1]$$

$$G(4,R) = [R_4,R_4][R_3,R_1,R_3][R_2,R_2,R_2]\cdot$$

$$[R_2,R_1,R_1,R_2][R_2,R_1,R_2,R_1][R_1,R_1,R_1,R_1,R_1]$$

etc.

6. STRUCTURE OF FOX MODULES

Let $\overline{F(k,R)} = F(k,R)/F(k+1,R)$, $k \ge 0$. Then $\overline{F(k,R)}$ can be regarded as a right F/R-module via conjugation in F. In this section we shall determine, for finite F/R, the structure of

$$U_k = \overline{F(k,R)} \otimes_{\mathbb{Z}} Q$$

as a right $Q(F/R)$-module. We call these modules the Fox modules of F/R. The structure of U_k, $k \ge 0$, has been completely described by Gupta & Passi (1981). Note that the case $k = 0$ is the relation module $R/R' \otimes_{\mathbb{Z}} Q$ studied in II.3. By Theorem 3.2 and Lemma 3.1,

$$F(k,R) = \sqrt{[R_k,R_k]\gamma_{k+1}(R)} = \sqrt{[R_{k+1},R_{k+1}]\gamma_{k+1}(R)}\ , \quad k \ge 1,$$

and it follows that for $k \ge 1$,

$$U_k = (\gamma_{k+1}(R)F(k+1,R)/F(k+1,R)) \otimes_{\mathbb{Z}} Q. \tag{17}$$

Let $1 \to R \overset{\theta}{\to} F \to G \to 1$ be a pre-abelian presentation of a finite group G with $F = \langle x_1,\ldots,x_m\rangle$, $m \ge 2$. Recall from Theorem II.3.6 that, as a Q-vector space, $U_0 = R/R' \otimes_{\mathbb{Z}} Q$ is spanned by the set

$$\{r_1^*R'\otimes 1,\ldots,r_m^*R'\otimes 1, r_{m+1}R'\otimes 1,\ldots,r_qR'\otimes 1\}$$

where $q = 1 + (m-1)|G|$, $r_i^* = r_i^{|G|}\eta_i$, $\eta_i \in R \cap F'$, $r_i = x_i^{e_i}\xi_i$, $i = 1,\dots,m$, $r_j = \xi_j$, $j = m+1,\dots,q$, $e_m|\dots|e_1 > 0$, $\xi_i \in F'$. Since $r_i^* \in R\backslash R \cap \gamma_2(F)$, it follows from the proof of Theorem 3.2 that, for $k \geq 1$, a basis for U_k consists of all elements of the form

$$\overline{b_{k+1}(r_1^*,\dots,r_m^*)} \otimes 1 \qquad \text{(Type I)}$$

and all elements of the form

$$\overline{[r_i, r_{i(1)}^*,\dots,r_{i(k)}^*]} \otimes 1 \qquad \text{(Type II)}$$

where $b_{k+1}(r_1^*,\dots,r_m^*)$ is a basic commutator of weight $k+1$ with entries from the set $\{r_1^*,\dots,r_m^*\}$ and $r_i \in \{r_{m+1},\dots,r_q\}$. By the Witt formula (Magnus et al. (1966), page 330), the number of elements of Type I (= rank $\gamma_{k+1}(F)/\gamma_{k+2}(F)$) is given by

$$\rho(k+1) = \frac{1}{k+1}\sum_{d|k+1} \mu(d)\, m^{(k+1)/d} \tag{18}$$

where

$$\mu(d) = \begin{cases} 1 & \text{if } d = 1 \\ 0 & \text{if } p^2|d,\ p \text{ prime} \\ (-1)^{\ell} & \text{if } d = p_1 \dots p_\ell,\ p_i\text{'s distinct primes} \end{cases}$$

is the Möbius function. The number of elements of Type II is clearly $m^k(q-m)$. We thus have proved,

6.1 *LEMMA*. For $k \geq 1$, the dimension of U_k is $\rho(k+1) + m^k(q-m)$ and the dimension of U_k^0, the fixed submodule of U_k, is $\rho(k+1)$. □

6.2 *COROLLARY*. If F/R is finite then $F(k,R)/F(k+1,R)$ is a free abelian group of rank $\rho(k+1) + m^k(q-m)$, where $m = \text{rank } F$, $q = \text{rank } R$. □

6.3 *LEMMA*. For $k \geq 0$, let χ_k denote the character of the Fox module U_k. Then

$$\chi_k(g) = \begin{cases} \rho(k+1) + m^k(1-m) & \text{for } g \neq 1 \\ \rho(k+1) + m^k(q-m) & \text{for } g = 1. \end{cases}$$

Proof. When $k = 0$, $u_0 = R/R' \otimes_{\mathbb{Z}} \mathbb{Q}$ and the result follows from Lemma II.3.2. Let $k \geq 1$. The action of $g = \theta(f) \in G$ on elements of Type I is given by

$$(\overline{b_{k+1}(r_1^*,\ldots,r_m^*)} \otimes 1)\cdot g$$

$$= \overline{b_{k+1}(r_1^{*f},\ldots,r_m^{*f})} \otimes 1$$

$$= \overline{b_{k+1}(r_1^*,\ldots,r_m^*)} \otimes 1$$

since $r_i^{*f} \equiv r_i^*$ (mod R'). Further, the action of g on elements of Type II is given by

$$(\overline{[r_j,r_{i(1)}^*,\ldots,r_{i(k)}^*]} \otimes 1)\cdot g$$

$$= \overline{[r_j^f,r_{i(1)}^*,\ldots,r_{i(k)}^*]} \otimes 1.$$

It follows that for each k-tuple $(r_{i(1)}^*,\ldots,r_{i(k)}^*)$, the subspace spanned by the Type II elements

$$\overline{[r_j,r_{i(1)}^*,\ldots,r_{i(k)}^*]} \otimes 1, \quad m+1 \le j \le q,$$

is isomorphic to the right G-module W (= subspace of U_0 spanned by the elements $r_jR' \otimes 1$, $m+1 \le j \le q$), under the map induced by

$$r_jR' \otimes 1 \to \overline{[r_j,r_{i(1)}^*,\ldots,r_{i(k)}^*]} \otimes 1.$$

Thus U_k is a direct sum of $\rho(k+1)$ copies of the trivial G-module Q and m^k copies of W. It follows that

$$\chi_k(g) = \rho(k+1) + m^k\psi_0(g),$$

where by II(16) and Theorem II.3.6, ψ_0 is the character of W given by

$$\psi_0(g) = \begin{cases} 1-m & \text{for } g \neq 1 \\ q-m & \text{for } g = 1. \end{cases}$$

Let χ denote the character of $\Delta_Q(G)$ (= QG(G-1)) as a right QG-module. Then

$$\chi(g) = \begin{cases} -1 & \text{for } g \neq 1 \\ |G| - 1 & \text{for } g = 1. \end{cases}$$

It follows that the character δ_k of V_k, where

$$V_k = \underbrace{Q \oplus \ldots \oplus Q}_{\leftarrow\rho(k+1)\rightarrow} + \underbrace{\Delta_Q(G) \oplus \ldots \oplus \Delta_Q(G)}_{\leftarrow m^k(m-1)\rightarrow}$$

is given by

$$\delta_k(g) = \begin{cases} \rho(k+1) + m^k(m-1)(-1) & \text{for } g \neq 1 \\ \rho(k+1) + m^k(m-1)(|G|-1) & \text{for } g = 1 \end{cases}$$

$$= \begin{cases} \rho(k+1) + m^k(1-m) & \text{for } g \neq 1 \\ \rho(k+1) + m^k(q-m) & \text{for } g = 1, \end{cases}$$

$(q = \text{rank } R)$

which, by Lemma 6.3, is the same as the character χ_k of U_k. Since the equivalence class of a representation is determined by its character, it follows that $U_k = V_k$. We have thus proved the following principal result of this section.

6.4 *THEOREM*. (Gupta & Passi 1981). For $k \geq 0$,

$$U_k = \underbrace{Q \oplus \ldots \oplus Q}_{\leftarrow\rho(k+1)\rightarrow} \oplus \underbrace{\Delta_Q(G) \oplus \ldots \oplus \Delta_Q(G)}_{\leftarrow m^k(m-1)\rightarrow},$$

where $m = \text{rank}(F)$ and $\rho(k+1)$ is given by (18). □

7. THE CENTRE OF $F/F(n,R)$, $n \geq 2$

Let $C/R' = \zeta(F/R')$ be the centre of F/R'. Then $C \leq R$ and C/R' is non-trivial if and only if F/R is finite. A complete characterization of C/R' is given in Chapter II, Section 1. In this section we obtain a complete description of the centre $\zeta(F/F(n,R))$ of $F/F(n,R)$, $n \geq 2$. We prove that $\zeta(F/F(n,R)) = \sqrt{\gamma_n(C)F(n,R)/F(n,R)}$, where C is as above. Throughout this section we shall assume that F/R is a finite group given by a pre-abelian presentation of the form (1).

We first prove

7.1 *LEMMA*. $\zeta(F/F(n,R)) \leq F(n-1,R)/F(n,R)$, $n \geq 1$.

Proof. By induction on $n \geq 1$. When $n = 1$, the proof follows from Proposition II.1.8. Let $n \geq 2$ and assume the result for $n-1$. Let $fF(n,R) \in \zeta(F/F(n,R))$. Then by the induction hypothesis, $f \in F(n-2,R)$ and $[f,w] \in F(n,R)$ for all $w \in F$. We choose $u \in R \cap F'\backslash R'$. Then $[u,f]-1 \equiv 0 \pmod{\mathfrak{r}\mathfrak{f}^n}$ implies

$$(u-1)(f-1) - (f-1)(u-1) \equiv 0 \pmod{\mathfrak{r}\mathfrak{f}^n}.$$

Thus, by Theorem 1.1, we have

$$\begin{aligned} 0 &= \alpha_{1,n+1}([u,f]-1) \\ &= \sum_{k=1}^{n+1} \alpha_{1,k}(u-1)\alpha_{k,n+1}(f-1) - \alpha_{1,k}(f-1)\alpha_{k,n+1}(u-1) \\ &= \alpha_{12}(u-1)\alpha_{2,n+1}(f-1), \end{aligned}$$

since $u-1 \in \mathfrak{f}^2$, $f-1 \in \mathfrak{r}\mathfrak{f}^{n-2} \le \mathfrak{f}^{n-1} \cap \mathfrak{r}$. Since $\alpha_{12}(u-1) \neq 0$ (Corollary I.1.10), it follows that $\alpha_{2,n+1}(f-1) = 0$ and hence $f-1 \in \mathfrak{f}^n$. Next, we choose $v \in R\backslash R \cap F'$. Then $[f,v]-1 \equiv 0 \pmod{\mathfrak{r}\mathfrak{f}^n}$ implies, as before,

$$\begin{aligned} 0 &= \alpha_{1,n+1}([f,v]-1) \\ &= \sum_{k=1}^{n+1} \alpha_{1,k}(f-1)\alpha_{k,n+1}(v-1) - \alpha_{1,k}(v-1)\alpha_{k,n+1}(f-1) \\ &= \alpha_{1,n}(f-1)\alpha_{n,n+1}(v-1), \end{aligned}$$

since $f-1 \in \mathfrak{f}^n \cap \mathfrak{r}\mathfrak{f}^{n-2}$. Since $\alpha_{n,n+1}(v-1) \neq 0$, we have $\alpha_{1,n}(f-1) = 0$ which, by Theorem 1.1, yields $f \in F(n-1,R)$ as desired. □

We can now prove our principal result.

7.2 *THEOREM*. (Gupta & Passi 1981). Let F/R be a finite group and let $C/R' = \zeta(F/R')$. Then, for $n \ge 2$, the centre of $F/F(n,R)$ is given by

$$\zeta(F/F(n,R)) = \sqrt{\gamma_n(C)F(n,R)/F(n,R)}\ .$$

Proof. Let $C^* = \langle r_1^*,\ldots,r_m^*\rangle \le R$, where r_i^* is as in Section 6. Then clearly $\gamma_n(C^*)F(n,R)/F(n,R)$ is a subgroup of $\zeta(F/F(n,R))$ and, being generated by Type I elements, has rank $\rho(n)$ given by (18). On the other hand, by Lemma 7.1, $\zeta(F/F(n,R)) \le F(n-1,R)/F(n,R)$ and hence

$$\zeta(F/F(n,R)) = \overline{F(n-1,R)}^{\,o},$$

the fixed $\mathbb{Z}(F/R)$-module of $\overline{F(n-1,R)} = F(n-1,R)/F(n,R)$. Since the dimension of U_{n-1}^{o}, the fixed submodule of

$U_{n-1} = \overline{F(n-1,R)} \otimes_{\mathbb{Z}} Q$, is also $\rho(n)$ (by Lemma 6.1) it follows that

$$\zeta(F/F(n,R)) = \sqrt{\gamma_n(C^*)/F(n,R)/F(n,R)}$$

Since $\sqrt{\gamma_n(C)F(n,R)/F(n,R)} \leq \overline{F(n-1,R)^{\circ}}$ the proof follows. □

Chapter IV

Dimension Subgroups

Let $\underline{\underline{f}} = \mathbb{Z}F(F-1)$ be the augmentation ideal of the free group ring $\mathbb{Z}F$. For any ideal $\underline{\underline{x}} \leq \underline{\underline{f}}$ and integer $n \geq 1$, the subgroup $D(n,\underline{\underline{x}}) = F \cap (1+\underline{\underline{x}}+\underline{\underline{f}}^n)$ will be called the n-th dimension subgroup of F relative to the ideal $\underline{\underline{x}}$. For $R \trianglelefteq F$, $\underline{\underline{r}} = \mathbb{Z}F(R-1)$, our ultimate goal is to achieve information about $D(n,\underline{\underline{r}})$ which is the same as $D(n,R)$ as defined in I.2.8. The motivation for the identification of $D(n,R)$ comes from an attempt to generalize to arbitrary integral group rings the fundamental theorem of free group rings. Observe that in the traditional notation, with $G = F/R$, the n-th dimension subgroup $D_n(G)$ $(= G \cap (1+\Delta^n(G))$ is isomorphic to $D(n,R)/R$ $(= D(n,\underline{\underline{r}})/R)$. We shall frequently use any of these notations without ambiguity.

1. SJOGREN'S THEOREM

In this section we shall prove Sjogren's Theorem that $D_{n+1}(G)/\gamma_{n+1}(G)$ has exponent dividing $C(n-1) = b(1)^{\binom{n-1}{1}} \cdots$ $\cdots b(n-1)^{\binom{n-1}{n-1}}$, $n \geq 2$, where $b(k) = \ell.c.m.\{1,\ldots,k\}$. Sjogren's proof which seems to have been inspired by Stallings (1975) is homological and uses the terminology of spectral sequences and the methods of Chen, Fox & Lyndon (1958). We present a simpler and more accessible proof of Sjogren's theorem. Much of the simplification in this exposition is due to Hartley (1982), Cliff & Hartley (1985) and Gupta (unpublished). We begin with some preliminary material required for the proof.

Let $n \geq 2$ be a fixed integer and let $\underline{\underline{A}}$ be the free (n+1)-truncated associative $\mathbb{Z}$-algebra generated by $\{a_1,a_2,\ldots\}$, i.e.,

$$\underline{\underline{A}} = A_0 \oplus A_1 \oplus \cdots \oplus A_n \tag{1}$$

where $A_0 = \mathbb{Z}$ and for $k \geq 1$, A_k is the homogeneous component of degree k with $\mathbb{Z}$-basis consisting of all monomials $c_1 \cdots c_k$,

$c_i \in \{a_1, a_2, \ldots\}$. Let

$$\underline{\underline{A}}^* = A_0^* \oplus A_1^* \oplus \ldots \oplus A_n^* \tag{2}$$

be the extended free associative $(n+1)$-truncated $\mathbb{Z}$-algebra generated by $\{a_1, a_2, \ldots, b_1, b_2, \ldots\}$. In what follows we shall introduce, for each $1 \le k \le n$, a special function $\phi_k\colon \underline{\underline{A}}^* \to \underline{\underline{A}}^* \otimes_{\mathbb{Z}} Q$ and establish some of its properties.

For each $1 \le \ell \le n$, $1 \le m \le n$, consider the polynomial

$$\frac{(-1)^{\ell+1}}{\ell} (a_1 \circ \ldots \circ a_m)^{\ell} \tag{3}$$

in $\underline{\underline{A}} \otimes_{\mathbb{Z}} Q$ where

$$a_1 \circ \ldots \circ a_m = (a_1+1) \ldots (a_m+1) - 1.$$

Let $\psi_\ell(a_1 \ldots a_m)$ denote the component of (3) whose monomials have positive degree in each of $a_1, \ldots, a_m$ (i.e., monomials with support $\{a_1, \ldots, a_m\}$). The polynomial $\psi_\ell(a_1 \ldots a_m)$, $1 \le m \le n$, defines, by linearity, the map

$$\psi_\ell\colon \underline{\underline{A}}^* \to \underline{\underline{A}}^* \otimes_{\mathbb{Z}} Q \tag{4}$$

defined by

$$\psi_\ell(c_1 \ldots c_m) = \psi_\ell(a_1 \ldots a_m)\Big|_{a_1=c_1, \ldots, a_m=c_m}, \tag{5}$$

the value of $\psi_\ell(a_1 \ldots a_m)$ at $a_i = c_i$, $c_i \in \{a_1, a_2, \ldots, b_1, b_2, \ldots\}$. For each $1 \le k \le n$, we define $\phi_k\colon \underline{\underline{A}}^* \to \frac{1}{b(k)} \underline{\underline{A}}^*$, $b(k) = \ell.\mathrm{c.m.}\{1, \ldots, k\}$, by

$$\phi_k = \sum_{\ell=1}^{k} \psi_\ell. \tag{6}$$

[Note: In view of (5) it is important to observe that the definitions of ψ_ℓ and ϕ_k are independent of the rank of $\underline{\underline{A}}$ and that $\psi_\ell(\underline{\underline{A}}^*)$, $\phi_k(\underline{\underline{A}}^*)$ are defined using a free algebra of rank n with $\ell, k \le n$.]

We now establish some of the properties of ϕ_k.

1.1 *LEMMA*. If $u \in A_k^*$ then $\phi_k(u) = u' + u''$, where u' is a Lie element in $\frac{1}{b(k)} A_k^*$ and $u'' \in \frac{1}{b(k)} (A_{k+1}^* \oplus \ldots \oplus A_n^*)$.

Proof. From the algebra of power series we have

$$e^{a_1} \ldots e^{a_k} = e^{\alpha} ,$$

where

$$\alpha = \log_e(1+v) = \sum_{\ell=1}^{\infty} \frac{(-1)^{\ell+1}}{\ell} v^{\ell}$$

and

$$v = e^{a_1} \ldots e^{a_k} - 1$$

$$= (1+a_1+\frac{1}{2!} a_1^2 + \ldots) \ldots (1+a_k+ \frac{1}{2!} a_k^2 + \ldots) - 1.$$

By the Baker-Cambell-Hausdorff formula (see, for instance, Magnus, Karrass & Solitar (1966), page 368) α is a Lie element in $\underset{=}{A} \otimes_{\mathbb{Z}} Q$ and its homogeneous component α_k of degree k is a Lie element in $A_k \otimes_{\mathbb{Z}} Q$. On the other hand, α_k is the homogeneous component of degree k of

$$\sum_{\ell=1}^{k} \frac{(-1)^{\ell+1}}{\ell} ((1+a_1) \ldots (1+a_k) - 1)^{\ell}.$$

It follows that $\phi_k(a_1 \ldots a_k) = w'+w''$, where w' is a Lie element in $\frac{1}{b(k)} A_k$ and $w'' \in \frac{1}{b(k)} (A_{k+1} \oplus \ldots \oplus A_n)$. The proof of the lemma follows by specialization. □

1.2 *LEMMA*. Let $u \in A_m^*$ be a $\mathbb{Z}$-linear sum of monomials of the form $a_{i(1)} \ldots a_{i(t-1)} b_{i(t)} a_{i(t+1)} \ldots a_{i(m)}$, $1 \le t \le m$, $b_{i(t)} \in \{b_1, b_2, \ldots\}$ and $a_{i(j)} \in \{a_1, a_2, \ldots\}$. Let $\delta: \underset{=}{A}^* \to \underset{=}{A}$ be an algebra homomorphism given by: $a_i \to a_i$, $i = 1,2,\ldots$; $b_j \to a_{j(1)} \circ \ldots \circ a_{j(q_j)}$, $j = 1,2,\ldots$. Then $\delta\phi_k(u) = \phi_k\delta(u)$ for all $1 \le k \le n$.

Proof. (Cliff & Hartley 1985). For each $q \ge 1$ let $c_1, \ldots, c_q$ be distinct elements of $\{a_{m+1}, a_{m+2}, \ldots\}$. For $1 \le t \le m$, let $\rho_t: \underset{=}{A} \to \underset{=}{A}$ be an algebra homomorphism given by: $a_i \to a_i$, $i \ne t$; $a_t \to c_1 \circ \ldots \circ c_q$. Then it clearly suffices to prove that $\psi_{\ell}\rho_t(a_1 \ldots a_m) = \rho_t\psi_{\ell}(a_1 \ldots a_m)$ for all $1 \le \ell \le n$. Indeed, we have in turn

$$\psi_\ell \rho_t(a_1 \dots a_m) = \psi_\ell(a_1 \dots a_{t-1}(c_1 \circ \dots \circ c_q)a_{t+1} \dots a_m)$$
$$= \sum_J \psi_\ell(a_1 \dots a_{t-1} c_{i(1)} \dots c_{i(j)} a_{t+1} \dots a_m),$$

where J runs through all non-empty subsequences $(c_{i(1)}, \dots, c_{i(j)})$ of $(c_1, \dots, c_q)$;

$$\psi_\ell \rho_t(a_1 \dots a_m) = \sum_J \text{component of } \frac{(-1)^{\ell+1}}{\ell} (a_1 \circ \dots \circ a_{t-1} \circ c_{i(1)} \circ \dots \circ c_{i(j)} \circ a_{t+1} \circ \dots \circ a_m)^\ell$$

with support $\{a_1, \dots, a_{t-1}, c_{i(1)}, \dots, c_{i(j)}, a_{t+1}, \dots, a_m\}$;

$$\psi_\ell \rho_t(a_1 \dots a_m) = \sum_J \text{component of } \frac{(-1)^{\ell+1}}{\ell} (a_1 \circ \dots \circ a_{t-1} \circ c_1 \circ \dots \circ c_q \circ a_{t+1} \circ \dots \circ a_m)^\ell$$

with support $\{a_1, \dots, a_{t-1}, c_{i(1)}, \dots, c_{i(j)}, a_{t+1}, \dots, a_m\}$;

$$\psi_\ell \rho_t(a_1 \dots a_m) = \text{component of } \frac{(-1)^{\ell+1}}{\ell} (a_1 \circ \dots \circ a_{t-1} \circ c_1 \circ \dots \circ c_q \circ a_{t+1} \circ \dots \circ a_m)^\ell$$

with support $\{a_1, \dots, a_{t-1},\ c_1 \circ \dots \circ c_q,\ a_{t+1}, \dots, a_q\}$;

$$\psi_\ell \rho_t(a_1 \dots a_m) = \rho_t(\text{component of } \frac{(-1)^{\ell+1}}{\ell} (a_1 \circ \dots \circ a_{t-1} \circ a_t \circ a_{t+1} \circ \dots \circ a_m)^\ell$$

with support $\{a_1, \dots, a_{t-1}, a_t, a_{t+1}, \dots, a_m\}$;

$$\psi_\ell \rho_t(a_1 \dots a_m) = \rho_t \psi_\ell(a_1 \dots a_m). \qquad \square$$

1.3 *LEMMA*. If $u \in A_m^*$ is a Lie element then, for all $1 \leq k \leq n$, $\phi_k(u) = u+u'$, where $u' \in \frac{1}{b(k)}(A_{m+1}^* \oplus \dots \oplus A_n^*)$. In particular, $\phi_k(u) = u$ for all Lie elements $u \in A_n^*$.

Proof. (Gupta (unpublished)). Let $u = ((a_1, \dots, a_m))$ be a (left normed) Lie element of degree m. Since $\psi_1(a_1 \dots a_m) = a_1 \dots a_m$, it follows that $\psi_1(u) = u$. Thus it suffices to prove that for $\ell \geq 2$, $\psi_\ell(u) \equiv 0 \mod (A_{m+1} \oplus \dots \oplus A_n) \otimes_{\mathbb{Z}} \mathbb{Q}$. When $m = 2$, we need only consider the case $\ell = 2$. We have

$$\begin{aligned}\psi_2((a_1,a_2)) &= \psi_2(a_1a_2 - a_2a_1)\\ &= \psi_2(a_1a_2) - \psi_2(a_2a_1)\\ &\equiv -\tfrac{1}{2}(a_1a_2 + a_2a_1) + \tfrac{1}{2}(a_2a_1 + a_1a_2)\\ &\equiv 0 \bmod (A_3 \oplus \ldots \oplus A_n) \otimes Q.\end{aligned}$$

For the inductive step assume $\psi_\ell((a_1,\ldots,a_m)) \in (A_{m+1} \oplus \ldots \oplus A_n) \otimes_{\mathbb{Z}} Q$ for all $\ell \geq 2$. Then

$$\psi_\ell((a_1,\ldots,a_{m+1})) = \psi_\ell\rho((a_2,\ldots,a_{m+1})),$$

where $\rho = \rho_1-\rho_2$, $\rho_1\colon a_i \to a_i$, $i \neq 2$, $a_2 \to a_1 \circ a_2$; and $\rho_2\colon a_i \to a_i$, $i \neq 2$, $a_2 \to a_2 \circ a_1$. By Lemma 1.2, $\psi_\ell\rho((a_2,\ldots,a_{m+1})) = \rho\psi_\ell((a_2,\ldots,a_{m+1})) = \rho(v)$, where by the induction hypothesis $v \in (A_{m+1} \oplus \ldots \oplus A_n) \otimes_{\mathbb{Z}} Q$ and $\text{support}(v) = \{a_2,\ldots,a_{m+1}\}$. Thus it suffices to prove if w is a monomial of length $m+1$ with support $\{a_2,\ldots,a_{m+1}\}$ then $\rho(w) \in A_{m+2} \oplus \ldots \oplus A_n$. Indeed, if $w = w_1a_2w_2$ and a_2 does not occur in w_1, w_2, then $\rho(w) = w_1a_1a_2w_2 - w_1a_2a_1w_2 \in A_{m+2}$. On the other hand, if $w = w_1a_2w_2a_2w_3$ then $\rho(w) = w_1(a_1 \circ a_2)w_2(a_1 \circ a_2)w_3 - w_1(a_2 \circ a_1)w_2(a_2 \circ a_1)w_3 \in A_{m+2} \oplus A_{m+3}$. This completes the proof of the lemma. □

At this stage it is convenient to include, for future reference, the following lemma which can also be deduced from a more general result of Hartley (1985).

1.4 *LEMMA*. Let $\underset{\sim}{A}$ denote the ideal of $\underline{\underline{A}}^*$ generated by all Lie elements of the form $((a_i,a_j))$. If $z \in \underset{\sim}{A}^2$ then $\phi_k(z) \in \underset{\sim}{A}^2 \otimes_{\mathbb{Z}} Q$ for all $k \geq 1$.

Proof. It clearly suffices to prove that $\phi_k(z) \in \underset{\sim}{A}^2 \otimes_{\mathbb{Z}} Q$ with

$$z = z_1((c_1,c_2))\,z_2((c_3,c_4))\,z_3$$

where the z_i are monomials in $\{a_1,a_2,\ldots,b_1,b_2,\ldots\}$ and $c_1,\ldots,c_4$ are distinct elements of $\{a_1,a_2,\ldots\}$ which are disjoint from the supports of the elements z_i. Indeed, let $b_\alpha, b_\beta \in \{b_1,b_2,\ldots\}$ be distinct from the supports of the elements z_i and put $z' = z_1b_\alpha z_2b_\beta z_3 \in \underline{\underline{A}}^*$. Then

$$z = (\lambda_1-\lambda_2)(\rho_1-\rho_2)(z')$$

where $\lambda_1, \lambda_2, \rho_1, \rho_2$ are algebra homomorphisms $\underline{\underline{A}}^* \to \underline{\underline{A}}^*$ defined by

$$\lambda_1 : b_\alpha \to c_1 \circ c_2,\ c_i \to c_i,\ c_i \neq b_\alpha$$

$$\lambda_2 : b_\alpha \to c_2 \circ c_1,\ c_i \to c_i,\ c_i \neq b_\alpha$$

$$\lambda_3 : b_\beta \to c_3 \circ c_4,\ c_i \to c_i,\ c_i \neq b_\beta$$

$$\lambda_4 : b_\beta \to c_4 \circ c_3,\ c_i \to c_i,\ c_i \neq b_\beta .$$

Then by Lemma 1.2 we have

$$\begin{aligned}\phi_k(z) &= \phi_k(\lambda_1-\lambda_2)(\rho_1-\rho_2)(z')\\ &= (\lambda_1-\lambda_2)(\rho_1-\rho_2)\phi_k(z')\end{aligned}$$

where $\phi_k(z')$, by definition, is a sum of monomials $u \in \underline{\underline{A}}^* \otimes_{\mathbb{Z}} Q$ with positive degree in each of b_α and b_β. Thus it suffices to prove that $(\lambda_1-\lambda_2)(\rho_1-\rho_2)(u)$ belongs to $\underset{\sim}{A}^2 \otimes_{\mathbb{Z}} Q$. For this we need the following formula which can be directly verified:

$$\begin{aligned}&u_1(c_i \circ c_j)\ldots u_t(c_i \circ c_j)u_{t+1}\\ &\qquad\qquad - u_1(c_j \circ c_i)\ldots u_t(c_j \circ c_i)u_{t+1}\\ &= \sum_{k=1}^{t} u_1(c_i \circ c_j)\ldots u_{k-1}(c_i \circ c_j)u_k((c_i,c_j))\\ &\qquad\qquad u_{k+1}(c_j \circ c_i)\ldots u_t(c_j \circ c_i)u_{t+1}.\end{aligned}$$

Using this in succession yields firstly, that $(\rho_1-\rho_2)(u)$ is a sum of elements $u' \in \underset{\sim}{A} \otimes_{\mathbb{Z}} Q$ where u' has positive degree in b_α; and secondly, that $(\lambda_1-\lambda_2)(u')$ lies in $\underset{\sim}{A}^2 \otimes_{\mathbb{Z}} Q$ as was to be proved. □

We shall now deduce Sjogren's key lemma for the dimension subgroups of $G = F/R$. We choose a positive countable presentation of G of the form

$$G = \langle x_1, x_2, \ldots ;\ r_1, r_2, \ldots \rangle , \tag{7}$$

where $r_i = x_{i(1)} \cdots x_{i(\ell_i)}$ is a semigroup word for all i. [For instance, $G = \langle x_1, x_2, \ldots ;\ r_1, r_2, \ldots \rangle = \langle x_1, x_2, \ldots, y_1, y_2, \ldots ;\ r_1, r_2, \ldots, x_1y_1, x_2y_2, \ldots \rangle$ can be reduced to a positive presentation by Tietze transformations.] Then $G \cong F/R$ where $F = \langle x_1, x_2, \ldots \rangle$ is free of finite or countable infinite rank and $R = \langle r_1, r_2, \ldots \rangle^F$ is the normal closure of the relators $r_1, r_2, \ldots$ For each r_i we choose a symbol y_i and consider the free group

$$F^* = \langle x_1, x_2, \ldots, y_1, y_2, \ldots \rangle .$$

Let $n \geq 2$ be fixed and let

$$\underline{A} = A_0 \oplus A_1 \oplus \ldots \oplus A_n; \ \underline{A}^* = A_0^* \oplus A_1^* \oplus \ldots \oplus A_n^*$$

be the $(n+1)$-truncated $\mathbb{Z}$-algebras on $\{a_1, a_2, \ldots\}$ and $\{a_1, a_2, \ldots, b_1, b_2, \ldots\}$ respectively, where each a_i corresponds to x_i and b_i corresponds to y_i in F^*. Let $\mathbb{Z}F$ and $\mathbb{Z}F^*$ be the free group rings with augmentation ideals $\underline{\mathfrak{f}}$ and $\underline{\mathfrak{f}}^*$ respectively. Then we have homomorphisms

$$\theta\colon \mathbb{Z}F/\underline{\mathfrak{f}}^{n+1} \to \underline{A}; \quad \theta^*\colon \mathbb{Z}F^*/\underline{\mathfrak{f}}^{*n+1} \to \underline{A}^*$$

given by $x_i + \underline{\mathfrak{f}}^{n+1} \to a_i + 1$; and $x_i + \underline{\mathfrak{f}}^{*n+1} \to a_i + 1$, $y_i + \underline{\mathfrak{f}}^{*n+1} \to b_i + 1$ respectively. We thus have the following commutative diagram

$$\begin{array}{ccc} \mathbb{Z}F^*/\underline{\mathfrak{f}}^{*n+1} & \xrightarrow{\ \beta\ } & \mathbb{Z}F/\underline{\mathfrak{f}}^{n+1} \\ \big\downarrow{\scriptstyle \theta^*} & & \big\downarrow{\scriptstyle \theta} \\ \underline{A}^* & \xrightarrow{\ \delta\ } & \underline{A} \end{array} \tag{8}$$

where β is given by $x_i \to x_i$, $y_j \to r_j$ and δ is given by $a_i \to a_i$, $b_j \to \theta(r_j)$ [observe also that, for $r_i = x_{i(1)} \cdots x_{i(\ell_i)}$, $\theta(r_i) = a_{i(1)} \circ \ldots \circ a_{i(\ell_i)}$]. Define $\underline{\mathfrak{r}}(1) = \underline{\mathfrak{r}} = \mathbb{Z}F(R-1) \leq \underline{\mathfrak{f}}$, and, more generally,

$$\underline{\mathfrak{r}}(k) = \sum_{i+j=k-1} \underline{\mathfrak{f}}^i \underline{\mathfrak{r}} \underline{\mathfrak{f}}^j, \quad k \geq 1,\ i,j \geq 0.$$

Also, set $R(1) = R$ and, more generally,

$$R(k) = [R, \underbrace{F, \ldots, F}_{k-1}], \quad k \geq 1.$$

[Note that $R(k) - 1 \leq \underline{\mathfrak{r}}(k)$.]

We now state and prove Sjogren's key lemma.

1.5(A) *LEMMA*. (Sjogren 1979). Let $w \in \gamma_n(F)$, $n \geq 2$, be such that $w - 1 \in \underline{\mathfrak{r}}(k) + \underline{\mathfrak{f}}^{n+1}$ for some $2 \leq k \leq n$. Then $w^{b(k)} - 1 \equiv f_k - 1 \bmod (\underline{\mathfrak{r}}(k+1) + \underline{\mathfrak{f}}^{n+1})$ for some $f_k \in R(k)$, where $b(k) = \ell.\text{c.m.}\{1, \ldots, k\}$.

Proof. Let $w - 1 \equiv u \bmod \underline{\mathfrak{f}}^{n+1}$, $u \in \underline{\mathfrak{r}}(k)$. Then we may write

$$u \equiv u_k + u_{k+1} + \dots + u_n \pmod{\underline{\underline{\mathfrak{f}}}^{n+1}}$$

where u_q, $k \le q \le n$, is a $\mathbb{Z}$-linear sum of elements of the form

$$(z_1-1)\dots(z_{t-1}-1)(r-1)(z_{t+1}-1)\dots(z_q-1)$$

with $z_i \in \{x_1,x_2,\dots\}$, $r \in \{r_1,r_2,..\}$. Thus denoting $\bar{z} = z+\underline{\underline{\mathfrak{f}}}^{n+1}$, $z \in \mathbb{Z}F$, we have by (8)

$$\theta(\overline{w-1}) = \theta(\bar{u}_k) + \dots + \theta(\bar{u}_n)$$

$$= \delta(v_k) + \dots + \delta(v_n), \tag{9}$$

where $\theta(\overline{w-1})$ is a Lie element in A_n and $v_q \in A_q^*$ is a $\mathbb{Z}$-linear sum of monomials of the form

$$c_1 \dots c_{t-1}bc_{t+1} \dots c_q$$

with $c_i \in \{a_1,a_2,\dots\}$, $b \in \{b_1,b_2,\dots\}$.

Since $\theta(\overline{w-1})$ is a Lie element in $A_n < A_n^*$, by Lemma 1.3 it follows that $\phi_k(\theta(\overline{w-1})) = \theta(\overline{w-1})$ for all k. Thus by Lemma 1.2, (9) yields

$$\theta(\overline{w-1}) = \phi_k\delta(v_k) + \dots + \phi_k\delta(v_n)$$

$$= \delta\phi_k(v_k) + \dots + \delta\phi_k(v_n).$$

By Lemma 1.1, $\phi_k(v_k) = v_k'+v_k''$, where v_k' is a Lie element in $\frac{1}{b(k)} A_k^*$ with one b-entry in each component and $v_k'' \in \frac{1}{b(k)}(A_{k+1}^* \oplus \dots \oplus A_n^*)$ is a sum of monomials with at least one b-entry. Also, for $q \ge k+1$, $\phi_k(v_q) \in \frac{1}{b(k)} (A_{k+1}^* \oplus \dots \oplus A_n^*)$ is a sum of monomials with at least one b-entry. Thus it follows by (8) that

$$b(k)\theta(\overline{w-1}) = \theta(\overline{f_k-1}) + \theta(\bar{u}')$$

for some $f_k \in R(k)$ and $u' \in \underline{\underline{\mathfrak{r}}}(k+1) + \underline{\underline{\mathfrak{f}}}^{n+1}$. This yields

$$\theta(\overline{w^{b(k)}-1}) = \theta(\overline{f_k-1}) + \theta(\bar{u}') ;$$

or equivalently,

$$w^{b(k)}-1 \equiv f_k-1 \bmod (\underline{\underline{\mathfrak{r}}}(k+1)+\underline{\underline{\mathfrak{f}}}^{n+1})$$

as was to be proved. □

When $k = n$, Lemma 1.5(A) admits a much sharper conclusion.

1.5(B) *LEMMA*. (Sjogren 1979). If $w \in \gamma_n(F)$, $n \geq 2$, be such that $w-1 \in \underline{\underline{r}}(n)+\underline{\underline{f}}^{n+1}$ then $w-1 \equiv g-1 \bmod \underline{\underline{f}}^{n+1}$ for some $g \in R(n)$. Equivalently, for $n \geq 2$, $F \cap (1+\underline{\underline{r}}(n)+\underline{\underline{f}}^{n+1}) = R(n)\gamma_{n+1}(F)$.

Proof. Since $w \in \gamma_n(F)$, $w-1$ is a Lie element of $\underline{\underline{f}}^n$ modulo $\underline{\underline{f}}^{n+1}$. Let $w-1 \equiv u \pmod{\underline{\underline{f}}^{n+1}}$ for some $u \in \underline{\underline{r}}(n)$. Then, by (8), $\theta(\overline{w-1}) = \theta(\bar{u})$ and $\theta(\overline{w-1})$ is a Lie element of degree n (in A_n). We may assume that $G \cong F/R$ is a finitely presented group and adopt a pre-abelian presentation

$$G = \langle x_1,\ldots,x_m;\ x_1^{e_1}\xi_1,\ldots,x_m^{e_m}\xi_m,\xi_{m+1},\ldots,\xi_q\rangle$$

where $\xi_i \in F'$ and $e_m|\ldots|e_1 \geq 0$.

Set $S = \langle x_1^{e_1},\ldots,x_m^{e_m},F'\rangle$, $\underline{\underline{s}} = \mathbb{Z}F(S-1)$ and $\underline{\underline{s}}(n) = \sum_{i+j=n-1} \underline{\underline{f}}^i\underline{\underline{s}}\underline{\underline{f}}^j$. Then, modulo $\underline{\underline{f}}^{n+1}$, u is an element of $\underline{\underline{s}}(n)$ and is a $\mathbb{Z}$-linear sum of elements of the form

$$(y_1-1)\ldots(y_{t-1}-1)(z-1)(y_{t+1}-1)\ldots(y_n-1),$$

where $y_i \in \{x_1,\ldots,x_m\}$ and $z \in \{x_1^{e_1},\ldots,x_m^{e_m}\}$.

Since the ideal $\underline{\underline{s}}(n)$ is invariant under substitutions $x_i \to 1$, $x_j \to x_j$, $j \neq i$, we may assume that u is a sum of elements of the above form involving each of $x_1,\ldots,x_m$ and hence also $e_m > 0$. Since $e_m|\ldots|e_1$, it follows that, modulo $\underline{\underline{f}}^{n+1}$, $u \equiv e_m v$, where v is a $\mathbb{Z}$-linear sum of elements of the form

$$(y_1-1)\ \ldots\ (y_n-1),$$

with $\{y_1,\ldots,y_n\} = \{x_1,\ldots,x_m\}$.

Thus $\theta(\overline{w-1}) = e_m\theta(\bar{v})$ is a Lie element of A_n and consequently, $\theta(\bar{v})$ is a Lie element of A_n. It follows that $\bar{v}$ is of the form $\overline{f-1}$ with $f = f(x_1,\ldots,x_m) \in \gamma_n(F)$. Thus u is of the form $f^{e_m}-1$ modulo $\underline{\underline{f}}^{n+1}$. Since x_m occurs in u, $f^{e_m} \equiv g \bmod \gamma_{n+1}(F)$ for some $g \in R(n)$ and, in turn, $w-1 \equiv g-1 \bmod \underline{\underline{f}}^{n+1}$ as was to be proved. □

We next prove a generalized version of a result due to Sjogren (1979) and Hartley (1982). Let $H = H_1 \geq H_2 \geq \ldots$ and $K = K_1 \geq K_2 \geq \ldots$ be series of normal subgroups of a group F (not necessarily free) and let $\{D_{k,\ell};\ 1 \leq k \leq \ell\}$ be a family of normal subgroups of F such that

(a) $D_{k,k+1} = H_k K_{k+1}$; (b) $H_k K_\ell \leq D_{k,\ell}$; and (c) $D_{k,\ell+1} \leq D_{k,\ell}$ for all $k < \ell$.

1.6 *THEOREM*. (Gupta 1985). For each $2 \leq k+m \leq n+1$, $k,m,n \geq 1$, let there exist a positive integer $a(k)$ (depending on k and n) such that

$$(K_{k+1} \cap D_{k,k+m+1})^{a(k)} \leq D_{k+1,k+m+1}H_k.$$

Then

$$D_{1,n+2}^{a(1,n+1)} \leq H_1 K_{n+2},$$

where $a(1,n+1) = a(1)^{\binom{n}{1}} \ldots a(n)^{\binom{n}{n}}$.

Proof. We prove by induction on $m \in \{1,\ldots,n\}$ that

$$D_{k,k+m+1}^{a(k,m+1)} \leq H_k K_{k+m+1},$$

where $a(k,m+1) = a(k)^{\binom{m}{1}} \ldots a(k+m-1)^{\binom{m}{m}}$; the proof of the theorem then follows taking $k = 1$ and $m = n$. When $m = 1$, we have

$$D_{k,k+2} \leq D_{k,k+1} = H_k K_{k+1},$$

by properties (c) and (a) above, and since $H_k \leq D_{k,k+2}$, it follows that

$$D_{k,k+2} \leq (K_{k+1} \cap D_{k,k+2})H_k.$$

Thus, by hypothesis,

$$\begin{aligned} D_{k,k+2}^{a(k)} &\leq D_{k+1,k+2}H_k \\ &= H_{k+1}K_{k+2}H_k \\ &= H_k K_{k+2}, \end{aligned}$$

as required, since $a(k) = a(k,2)$. For the inductive step, let $m \geq 2$ and assume the result for $m-1$. By definition

$$D_{k,k+m+1} \le D_{k,k+(m-1)+1}$$

and hence by the induction hypothesis,

$$D_{k,k+m+1}^{a(k,m)} \le (H_k K_{k+m}) \cap D_{k,k+m+1}$$

$$\le (K_{k+m} \cap D_{k,k+m+1}) H_k ,$$

since $H_k \le D_{k,k+m+1}$. This, in turn, yields

$$D_{k,k+m+1}^{a(k,m)a(k)} \le (K_{k+m} \cap D_{k,k+m+1})^{a(k)} H_k$$

$$\le D_{k+1,k+m+1} H_k ,$$

by hypothesis. On the other hand, $D_{k+1,k+m+1} = D_{k+1,k+1+m}$ and by the induction hypothesis

$$D_{k+1,k+1+m}^{a(k+1,m)} \le H_{k+1} K_{k+1+m}.$$

Thus

$$D_{k,k+m+1}^{a(k,m)a(k)a(k+1,m)} \le H_{k+1} K_{k+m+1}$$

and it only remains to verify that $a(k,m+1) = a(k,m)a(k)a(k+1,m)$. Indeed, we have

$$a(k,m) = a(k)^{\binom{m-1}{1}} \cdots a(k+m-2)^{\binom{m-1}{m-1}}$$

and

$$a(k+1,m) = a(k+1)^{\binom{m-1}{1}} \cdots a(k+1+m-2)^{\binom{m-1}{m-1}} .$$

Thus

$$a(k,m)a(k)a(k+m)$$

$$= a(k)^{\binom{m-1}{0}+\binom{m-1}{1}} a(k+1)^{\binom{m-1}{1}+\binom{m-1}{2}} \cdots$$

$$\cdots a(k+m-2)^{\binom{m-1}{m-2}+\binom{m-1}{m-1}} a(k+m-1)$$

$$= a(k)^{\binom{m}{1}} a(k+1)^{\binom{m}{2}} \cdots a(k+m-2)^{\binom{m}{m-1}} a(k+m-1)^{\binom{m}{m}}$$

$$= a(k,m+1). \qquad \square$$

We can now prove our main result of this section.

1.7 *THEOREM.* (Sjogren 1979). For all groups G, $D_{n+1}(G)/\gamma_{n+1}(G)$ has exponent dividing $C(n-1) = b(1)^{\binom{n-1}{1}} \dots b(n-1)^{\binom{n-1}{n-1}}$, $n \geq 2$, where $b(k) = \ell.c.m.\{1,\dots,k\}$.

Proof. Let $G \cong F/R$ be given by a positive presentation of the form (7) and let $n \geq 2$ be fixed. Let $w \in D(n+1,R) = F \cap (1+\underline{\underline{r}}+\underline{\underline{f}}^{n+1})$. Then the proof consists in showing that $w^{C(n-1)} \in R\gamma_{n+1}(F)$. We proceed as follows: Set $H_k = R(k)$, $k \geq 1$, $K_\ell = \gamma_\ell(F)$, $\ell \geq 1$ and $D_{k,\ell} = D(\ell,\underline{\underline{r}}(k)) = F \cap (1+\underline{\underline{r}}(k)+\underline{\underline{f}}^\ell)$. Then $D_{k,k+1} = H_kK_{k+1}$ (Lemma 1.5(B)), $H_kK_\ell \leq D_{k,\ell}$ and $D_{k,\ell+1} \leq D_{k,\ell}$ for all $k < \ell$. In addition, by Lemma 1.5(A) $(K_{k+1} \cap D_{k,k+m+1})^{b(k)} \leq D_{k+1,k+m+1}H_k$ for all $2 \leq k+m \leq n+1$, $k,m,n \geq 1$. It follows by Theorem 1.6 that $D_{1,n+1}/H_1K_{n+1}$ has exponent dividing $b(1)^{\binom{n-1}{1}} \dots b(n-1)^{\binom{n-1}{n-1}} = C(n-1)$ and hence $D(n+1,\underline{\underline{r}}(1))/R(1)\gamma_{n+1}(F)$ has exponent dividing $C(n-1)$. Since $\underline{\underline{r}}(1) = \underline{\underline{r}}$ and $R(1) = R$, we have the desired result. □

If p is prime then $C(p-1)$ is coprime to p and we deduce,

1.8 *COROLLARY.* (Sjogren 1979). If G is a p-group then $D_n(G) = \gamma_n(G)$ for $n \leq p+1$. □

[This improves an earlier result due to Moran (1970) that $D_n(G) = \gamma_n(G)$ for $n \leq p$.]

Several earlier known results follow from Sjogren's Theorem 1.7.

1.9 *COROLLARY.* (Hall-Jennings). If all the lower central factors $\gamma_k(G)/\gamma_{k+1}(G)$ of a group G are torsion-free then $D_n(G) = \gamma_n(G)$ for all $n \geq 1$. □

[See also Quillen (1968).]

1.10 *COROLLARY.* (Higman-Rees). For every group G, $D_3(G) = \gamma_3(G)$. □

[Other proofs of Corollary 1.10 have been given by Passi (1968), Hoare (1969), Sandling (1972), Bachmann & Grunenfelder (1972).

See end of Theorem 3.2 for an alternative proof due to Gupta (1982).]

1.11 *COROLLARY*. (Losey 1974). For every group G, $D_4^2(G) \leq \gamma_4(G)$. □

[Other proofs are due to Tahara (1977), Passi (1979). See also Remark 4.8.]

1.12 *COROLLARY*. (Passi 1968). If G is a p-group, p odd, then $D_4(G) = \gamma_4(G)$. □

1.13 *REMARKS*. The bound $C(n-1)$ for the exponent of $D_{n+1}(G)/\gamma_{n+1}(G)$ is unlikely to be best possible. If G is assumed to be metabelian then Gupta (1984) has proved that $C(n-1)$ can be replaced by a smaller integer $2 \cdot C(n-1)^*$ where $C(n-1)^* = b(1)\ldots b(n-1)$. (See Theorem 4.6.) Further, for metabelian p-groups, p odd, Gupta & Tahara (1985) have proved that $D_n(G) = \gamma_n(G)$ for $n \leq p+2$. For an arbitrary group G, Tahara (1981) has proved that $D_5(G)/\gamma_5(G)$ has exponent dividing 6. Sjogren's theorem also yields a result about Schur muliplicator (Passi & Vermani (1983)).

2. A COUNTER-EXAMPLE DUE TO RIPS

It had long been conjectured that for any group G, $D_n(G) = \gamma_n(G)$ for all $n \geq 1$. If for some $n \geq 1$, $D_n(G) \neq \gamma_n(G)$ for some group G then it is not difficult to see that there is a finite p-group G for which $D_n(G) \neq \gamma_n(G)$ (see, for instance, Passi (1979), page 59). Thus for the purpose of counter-examples it is enough to concentrate on finite p-groups. In particular, by Corollary 1.12, it follows that if $D_4(G) \neq \gamma_4(G)$ then there must be a finite 2-group G for which $D_4(G) \neq \gamma_4(G)$. Such a group was first constructed by Rips in 1972 providing a counter-example to the dimension subgroup conjecture and is regarded as a major contribution to the dimension subgroup problem. In the following example we construct a 2-group G, based on Rips' example, with $D_4(G) \neq \gamma_4(G)$. The present exposition is due to Gupta & Passi (unpublished). A discussion with Dr. M.F. Newman is gratefully acknowledged.

2.1 *EXAMPLE*. (cf. Rips 1972).

Let $F = \langle x_1, x_2, x_3, x_4 \rangle$ be a free group. Set

$$R_1 = \gamma_4(F);$$

$$R_2 = \langle R_1,\ [x_i,x_j,x_k] \notin \langle\alpha,\beta,\gamma\rangle R_1, \quad \text{for all} \quad i,j,k\rangle$$

where $\alpha = [x_1,x_2,x_2]$, $\beta = [x_1,x_3,x_3]$, $\gamma = [x_1,x_4,x_4]$;

$$R_3 = \langle R_2,a_1,a_2,a_3\rangle,$$

where $a_1 = \alpha^4\beta^{-1}$, $a_2 = \beta^4\gamma^{-1}$, $a_3 = \gamma^4$;

$$R_4 = \langle R_3,b_1,\ldots,b_6\rangle,$$

where $b_1 = [x_1,x_2]^{64}\alpha^{32}$, $b_2 = [x_1,x_3]^{16}\beta^8$, $b_3 = [x_1,x_4]^4\gamma^2$, $b_4 = [x_2,x_3]^{16}\beta^{-1}$, $b_5 = [x_2,x_4]^4\alpha^{-2}$, $b_6 = [x_3,x_4]^4\beta^{-1}$; and finally,

$$R = R_5 = \langle R_4,c_1,c_2,c_3,c_4\rangle,$$

where

$$c_1 = x_1^{64}[x_1,x_2]^{32},$$

$$c_2 = x_2^{64}[x_1,x_3]^{-4}[x_1,x_4]^{-2},$$

$$c_3 = x_3^{16}[x_1,x_2]^4[x_1,x_4]^{-1},$$

$$c_4 = x_4^4[x_1,x_2]^2[x_1,x_3].$$

We first verify that $[R_{i+1},F] \le R_i$ for $i = 1,\ldots,4$. For $i = 1,2,3$ this is clear. To see that $[R_5,F] \le R_4$ we verify each of the 16 instances $[c_i,x_j]$, $i,j = 1,\ldots,4$, separately. Indeed, modulo R_4, we have

$$[c_1,x_1] \equiv 1;$$

$$[c_1,x_2] \equiv [x_1^{64},x_2][x_1,x_2,x_2]^{32} \equiv [x_1,x_2]^{64}\alpha^{32} \equiv b_1 \equiv 1;$$

$$[c_1,x_3] \equiv [x_1,x_3]^{64} \equiv b_2^4\beta^{-32} \equiv \gamma^{-8} \equiv 1;$$

$$[c_1,x_4] \equiv [x_1,x_4]^{64} \equiv \gamma^{-32} \equiv 1;$$

$$[c_2,x_1] \equiv [x_2^{64},x_1] \equiv [x_2,x_1]^{64}[x_2,x_1,x_2]^{\binom{64}{2}}$$

$$\equiv [x_1,x_2]^{-64}\alpha^{32} \equiv 1;$$

$$[c_2,x_2] \equiv 1;$$

$$[c_2,x_3] \equiv [x_2,x_3]^{64}\beta^{-4} \equiv 1;$$

$$[c_2,x_4] \equiv [x_2,x_4]^{64}\gamma^{-2} \equiv [x_2,x_4]^{64}\alpha^{-32} \equiv 1;$$

$$[c_3,x_1] \equiv [x_3^{16},x_1] \equiv [x_3,x_1]^{16}[x_3,x_1,x_3]^{\binom{16}{2}}$$

$$\equiv [x_1,x_3]^{-16}\beta^{-8} \equiv 1;$$

$$[c_3,x_2] \equiv [x_3,x_2]^{16}[x_1,x_2,x_2]^4 \equiv [x_2,x_3]^{-16}\beta \equiv 1;$$

$$[c_3,x_3] \equiv 1;$$

$$[c_3,x_4] \equiv [x_3,x_4]^{16}[x_1,x_4,x_4]^{-1} \equiv [x_3,x_4]^{16}\gamma^{-1}$$

$$\equiv [x_3,x_4]^{16}\beta^{-4} \equiv 1;$$

$$[c_4,x_1] \equiv [x_1,x_4^4]^{-1} \equiv [x_1,x_4]^{-4}[x_1,x_4,x_4]^{-\binom{4}{2}}$$

$$\equiv [x_1,x_4]^{-4}\gamma^{-2} \equiv 1;$$

$$[c_4,x_2] \equiv [x_2,x_4]^{-4}[x_1,x_2,x_2]^2 \equiv [x_2,x_4]^{-4}\alpha^2 \equiv 1;$$

$$[c_4,x_3] \equiv [x_3,x_4]^{-4}[x_1,x_3,x_3] \equiv [x_3,x_4]^{-4}\beta \equiv 1;$$

$$[c_4,x_4] \equiv 1.$$

Since $[R_{i+1},F] \le R_i$ and $R_1 \le R_2 \le R_3 \le R_4 \le R_5 = R$, it follows that each R_i is a normal subgroup of F. Further, $F/R = G$ is a finite nilpotent group of class precisely 3 and order $2^6 2^{20} 2^{18} = 2^{44}$. This completes the construction of G. □

2.2 $\underline{D_4(G) \ne \gamma_4(G)}$

Let

$$w = [x_2^{64},x_3]^2[x_2^{64},x_4][x_3^{16},x_4]^2.$$

Then

$$w \equiv [x_2,x_3]^{128}[x_2,x_4]^{64}[x_3,x_4]^{32} \pmod{R_3}$$

$$\equiv \beta^8\alpha^{32}\beta^8 \pmod{R_4}$$

$$\equiv \alpha^{32} \pmod{R_4}$$

$$\not\equiv 1 \pmod{R},$$

since α is of order 64. Thus $w \notin \gamma_4(G)$ and in what follows we shall prove that $w-1 \in \underline{\underline{r}}+\underline{\underline{f}}^4$.

Let $S = \langle x_1^{64}, x_2^{64}, x_3^{16}, x_4^4, F' \rangle \trianglelefteq F$ and $\underline{\underline{s}} = \mathbb{Z}F(S-1)$. Further set $\underline{\underline{x}} = \underline{\underline{s}}\underline{\underline{f}}\underline{\underline{f}} + \underline{\underline{f}}\underline{\underline{s}}\underline{\underline{f}} + \underline{\underline{f}}\underline{\underline{f}}\underline{\underline{s}}$. Then by the definition of R_5 it follows that $\underline{\underline{x}} \leq \underline{\underline{r}} + \underline{\underline{f}}^4$. Now, expansion of $w-1$ modulo $(\underline{\underline{r}}+\underline{\underline{f}}^4)$ yields

$$\begin{aligned} w-1 \equiv\ & 128\{(x_2-1)(x_3-1) - (x_3-1)(x_2-1)\} \\ & + 64\{(x_2-1)(x_4-1) - (x_4-1)(x_2-1)\} \\ & + 32\{(x_3-1)(x_4-1) - (x_4-1)(x_3-1)\} \\ \equiv\ & (x_2-1)(y_2-1) + (x_3-1)(y_3-1) + (x_4-1)(y_4-1), \end{aligned}$$

where $y_2 = x_3^{128}x_4^{64}$, $y_3 = x_2^{-128}x_4^{32}$, $y_4 = x_2^{-64}x_3^{-32}$. If for some $\xi_i \in F'$, $y_i \equiv \xi_i^{e_i} \bmod R\,\gamma_3(F)$, with $e_2 = 64$, $e_3 = 16$ and $e_4 = 4$, then, modulo $\underline{\underline{f}}\underline{\underline{r}} + \underline{\underline{f}}^4$,

$$\begin{aligned} w-1 &\equiv 64(x_2-1)(\xi_2-1) + 16(x_3-1)(\xi_3-1) + 4(x_4-1)(\xi_4-1) \\ &\equiv (x_2^{64}-1)(\xi_2-1) + (x_3^{16}-1)(\xi_3-1) + (x_4^4-1)(\xi_4-1) \end{aligned}$$

which is an element of $\underline{\underline{x}}+\underline{\underline{f}}^4 \leq \underline{\underline{r}}+\underline{\underline{f}}^4$. Thus it suffices to verify that y_2, y_3, y_4 have the stated properties. We have

$$\begin{aligned} y_2 &= x_3^{128}x_4^{64} \\ &\equiv [x_1,x_2]^{-32}[x_1,x_4]^{8}[x_1,x_2]^{-32}[x_1,x_3]^{-16} \pmod{R_5} \\ &\equiv \alpha^{32}\beta^{8}\gamma^{-4} \pmod{R_4} \\ &\equiv 1 \ (R_3), \end{aligned}$$

so we may choose $\xi_2 = 1$, $e_2 = 64$;

$$\begin{aligned} y_3 &= x_2^{-128}x_4^{32} \\ &\equiv [x_1,x_3]^{-8}[x_1,x_4]^{-4}[x_1,x_2]^{-16}[x_1,x_3]^{-8} \pmod{R_5} \end{aligned}$$

$$\equiv \beta^8\gamma^2[x_1,x_2]^{-16} \pmod{R_4}$$

$$\equiv [x_1,x_2]^{-16} \bmod R\,\gamma_3(F),$$

so we may choose $\xi_3 = [x_1,x_2]^{-1}$, $e_3 = 16$; and finally,

$$y_4 = x_2^{-64}x_3^{-32}$$

$$\equiv [x_1,x_3]^{-4}[x_1,x_4]^{-2}[x_1,x_2]^8[x_1,x_4]^{-2} \pmod{R_5}$$

$$\equiv \gamma^2[x_1,x_1]^{-4}[x_1,x_2]^8 \bmod R_4$$

$$\equiv ([x_1,x_4]^{-1}[x_1,x_3]^{-1}[x_1,x_2]^2)^4 \bmod R\,\gamma_3(F),$$

and we choose $\xi_4 = [x_1,x_4]^{-1}[x_1,x_3]^{-1}[x_1,x_2]^2$, $e_4 = 4$. □

2.3 *REMARKS*. The above counter-example to the dimension subgroup conjecture is a metabelian group. Thus it is significant to study the dimension subgroups of metabelian groups. In the subsequent sections we shall seek a detailed analysis of the quotients $D_n(G)/\gamma_n(G)$ when G is assumed to be metabelian.

3. SOME PRELIMINARY RESULTS ON DIMENSION SUBGROUPS

Let $w \in D(n,R) = F \cap (1+\underline{\underline{r}}+\underline{\underline{f}}^n)$. Then $w-1 \in \underline{\underline{r}}+\underline{\underline{f}}^n$ and it follows that $w-1 \equiv u \bmod (\underline{\underline{f}}\underline{\underline{r}}+\underline{\underline{f}}^n)$, where $u = \sum_i n_i(r_i-1) \equiv (\prod_i r_i^{n_i}-1) \bmod \underline{\underline{f}}\underline{\underline{r}}$, $n_i \in \mathbb{Z}$, $r_i \in R$. Thus $w-1 \equiv r-1 \bmod (\underline{\underline{f}}\underline{\underline{r}}+\underline{\underline{f}}^n)$ where $r = \prod_i r_i^{n_i} \in R$. It follows that

$$wr^{-1} \in F \cap (1+\underline{\underline{f}}\underline{\underline{r}}+\underline{\underline{f}}^n) = D(n,\underline{\underline{f}}\underline{\underline{r}}).$$

If $D(n,\underline{\underline{f}}\underline{\underline{r}}) \leq R\,\gamma_n(F)$, then $wr^{-1} \in R\,\gamma_n(F)$ and hence $w \in R\,\gamma_n(F)$. Thus we have,

3.1 *PROPOSITION*. $D(n,R) = R\,\gamma_n(F)$ if and only if $D(n,\underline{\underline{f}}\underline{\underline{r}}) \leq R\,\gamma_n(F)$. □

For the remainder of this chapter we shall assume that $G = F/R$ is a finitely generated metabelian group given by a pre-abelian presentation

$$F/R = \langle x_1,\ldots,x_m;\ x_1^{e_1}\xi_1,\ldots,x_m^{e_m}\xi_m,\xi_{m+1},\ldots,F''\rangle, \tag{10}$$

with $e_m | \ldots | e_1 \geq 0$, $\xi_i \in F'$ for $i = 1,2,\ldots$.

Let $S = \langle x_1^{e_1}, \ldots, x_m^{e_m}, F' \rangle$ be a normal subgroup of the free group $F = \langle x_1, \ldots, x_m \rangle$. Then F/S is abelian with $S' \leq R$.

We regard F'/F'' as a right $\mathbb{Z}F$-module via conjugation in F. We, however, continue to write F'/F'' multiplicatively. For $w \in F'$, $f \in F$, $u,v \in \mathbb{Z}F$, we have

$$(wF'')^f = w^f F'' = (f^{-1}wf)F''$$

and

$$(wF'')^{u+v} = (wF'')^u (wF'')^v.$$

For convenience, while working modulo F'', we shall write w^u, $w \in F'$, $u \in \mathbb{Z}F$, to mean a pre-image in F' of $(wF'')^u$. Throughout we use left-normed commutator convention. The following congruences hold, modulo F'', for all $x,y,z \in F$,

$$[x,y,z][y,z,x][z,x,y] \equiv 1 \qquad \text{(Jacobi)}; \tag{11}$$

$$[x^n,y] \equiv [x,y]^{1+x+\ldots+x^{n-1}}, \quad n \geq 1; \tag{12}$$

$$[x,y]^{z-1} \equiv [x,y,z]. \tag{13}$$

We write

$$t(x_i,e_i) = \begin{cases} 1 + x_i + \ldots + x_i^{e_i-1}, & e_i \geq 1 \\ 0, & e_i = 0. \end{cases}$$

Then $(x_i-1)t(x_i,e_i) = x_i^{e_i}-1 \in \mathfrak{s} = \mathbb{Z}F(S-1)$, $1 \leq i \leq m$. We first prove,

3.2 *THEOREM*. (Gupta 1982). For all $n \geq 1$, modulo $[F',S]\gamma_{n+2}(F)$, $D(n+2,\mathfrak{f}\mathfrak{s})$ is generated by the elements

$$[x_i,x_j]^{t(x_i,e_i)a_{ij}}, \quad 1 \leq i < j \leq m,$$

where $a_{ij} = a_{ij}(x_j,\ldots,x_m) \in \mathbb{Z}F$ and $t(x_i,e_i)a_{ij} \in t(x_j,e_j)\mathbb{Z}F + \mathfrak{s} + \mathfrak{f}^n$.

Proof. Let $t(x_i,e_i)a_{ij} \in t(x_j,e_j)\mathbb{Z}F + \mathfrak{s} + \mathfrak{f}^n$. Then we have the following sequence of congruences,

$$[x_i,x_j]^{t(x_i,e_i)a_{ij}}-1 \equiv \{(x_i-1)(x_j-1) - (x_j-1)(x_i-1)\} \times t(x_i,e_i)a_{ij} \pmod{\mathfrak{f}\mathfrak{s}};$$

$$\equiv (x_i-1)(x_j-1)t(x_i,e_i)a_{ij} \pmod{\mathfrak{f}\mathfrak{s}},$$

since $(x_i-1)t(x_i,e_i) \in \mathfrak{s}$;

$$\equiv 0 \pmod{\mathfrak{f}\mathfrak{s} + \mathfrak{f}^{n+2}},$$

since $t(x_i,e_i)a_{ij} \in t(x_j,e_j)\mathbb{Z}F + \mathfrak{s} + \mathfrak{f}^n$.

Thus $[x_i,x_j]^{t(x_i,e_i)a_{ij}} \in D(n+2,\mathfrak{f}\mathfrak{s})$.

Conversely, let $w \in D(n+2,\mathfrak{f}\mathfrak{s})$, $n \geq 1$. Then $w \in F'$ and, modulo F'', we can express w as

$$w \equiv \prod_{1\leq i<j\leq m} [x_i,x_j]^{d_{ij}}, \quad d_{ij} = d_{ij}(x_1,\ldots,x_m) \in \mathbb{Z}F .$$

By Jacobi congruence we have, for $1 \leq k < i < j$,

$$[x_i,x_j]^{x_k-1} \equiv [x_i,x_j,x_k] \equiv [x_k,x_i]^{-(x_j-1)}[x_k,x_j]^{(x_i-1)}.$$

Using this, if necessary, we may rewrite w as above where, in addition, $d_{ij} = d_{ij}(x_i,\ldots,x_m)$. We re-organize w as

$$w \equiv w_1 \ldots w_{m-1},$$

where

$$w_i \equiv \prod_{j=i+1}^{m} [x_i,x_j]^{d_{ij}}. \tag{16}$$

Let θ_i: $\mathbb{Z}F \to \mathbb{Z}F$, $1 \leq i \leq m-1$, be the ring homomorphism given by $x_i \to 1$, $x_j \to x_j$, $j \neq i$. Then the ideals $\mathfrak{s}$ and $\mathfrak{f}$ are invariant under all θ_i's and it follows that since $w \in D(n+2,\mathfrak{f}\mathfrak{s})$, applying $\theta_1,\ldots,\theta_{m-1}$ successively gives $w_i \in D(n+2,\mathfrak{f}\mathfrak{s})$ for each $i = 1,2,\ldots,m-1$. Thus modulo $\mathfrak{f}\mathfrak{s}+\mathfrak{f}^{n+2}$, we have

$$\begin{aligned} 0 &\equiv w_i-1 \\ &\equiv \sum_{j=i+1}^{m} ([x_i,x_j]-1)d_{ij} \\ &\equiv \sum_{j=i+1}^{m} ((x_i-1)(x_j-1)d_{ij} - (x_j-1)(x_i-1)d_{ij}) \\ &= (x_i-1)\sum_{j=i+1}^{m} (x_j-1)d_{ij} - \sum_{j=i+1}^{m} (x_j-1)(x_i-1)d_{ij}. \end{aligned}$$

Since $\mathfrak{f}$ is a right $\mathbb{Z}F$-module with basis $\{x_i-1,\ i = 1,\dots,m\}$ (Remark I.1.11), it follows that

$$(x_i-1)d_{ij} \equiv 0 \bmod \mathfrak{s} + \mathfrak{f}^{n+1} \tag{17}$$

and

$$\sum_{j=i+1}^{m} (x_j-1)d_{ij} \equiv 0 \bmod \mathfrak{s} + \mathfrak{f}^{n+1}. \tag{18}$$

Now

$$\mathfrak{s} = \sum_{i=1}^{m} (x_i-1)t(x_i,e_i)\mathbb{Z}F + \mathfrak{a},$$

where $\mathfrak{a} = \mathbb{Z}F(F'-1)$ and

$$\mathfrak{s}+\mathfrak{f}^{n+1} = \sum_{i=1}^{m} (x_i-1)(t(x_i,e_i)\mathbb{Z}F+\mathfrak{f}^{n}) + \mathfrak{a}.$$

Since $\mathbb{Z}F/\mathfrak{a} \cong \mathbb{Z}(F/F')$ is an integral domain, (17) yields

$$d_{ij} \in t(x_i,e_i)\mathbb{Z}F + \mathfrak{f}^{n} + \mathfrak{a}$$

which on substituting in (16) gives, using (12),

$$w_i \equiv \prod_{j=i+1}^{m} [x_i^{e_i},x_j]^{a_{ij}} \bmod F''\gamma_{n+2}(F)$$

where $a_{ij} = a_{ij}(x_i,\dots,x_m) \in \mathbb{Z}F$.

Since $[x_i^{e_i},x_j]^{x_k-1} \equiv [x_i^{e_i},x_k]^{x_j-1} \bmod [F',S]$, by rewriting w_i, if necessary, we may assume that

$$w_i \equiv \prod_{j=i+1}^{m} [x_i^{e_i},x_j]^{a_{ij}} \bmod [F',S]\gamma_{n+2}(F)$$

where $a_{ij} = a_{ij}(x_j,\dots,x_m) \in \mathbb{Z}F$.

This yields,

$$w_i \equiv \prod_{j=i+1}^{m} [x_i,x_j]^{t(x_i,e_i)a_{ij}} \bmod [F',S]\gamma_{n+2}(F)$$

with $d_{ij} = t(x_i,e_i)a_{ij}$, $a_{ij} = a_{ij}(x_j,\dots,x_m) \in \mathbb{Z}F$.

From (18) it now follows that

$$d_{ij} \in t(x_j,e_j)\mathbb{Z}F + \mathfrak{s} + \mathfrak{f}^{n}.$$

This completes the proof of the theorem. □

[Since, by (10), $[x_i,x_j]^{t(x_i,e_i)} \in R\,\gamma_3(F)$, Theorem 3.2 yields $D(3,\mathfrak{f}\mathfrak{s}) \le R\,\gamma_3(F)$. Thus $D_3(G) = \gamma_3(G)$ for all G giving another proof of Corollary 1.10.]

If $S = F'$, then, in (10), $e_i = 0$ for all i and Theorem 3.2 yields the following important corollary.

3.3 *THEOREM*. (Gupta 1982). $F \cap (1+\mathfrak{f}\mathfrak{a}+\mathfrak{f}^{n+2}) = F''\,\gamma_{n+2}(F)$ where $\mathfrak{a} = \mathbb{Z}F(F'-1)$. □

3.4 *PROPOSITION*. If F/S is a finite abelian group of exponent e then $D(n+2,\mathfrak{f}\mathfrak{s})^e \le S'\,\gamma_{n+2}(F)$.

Proof. By Theorem 3.2, it suffices to prove that if $d_{ij} \in t(x_i,e_i)\mathbb{Z}F \cap (t(x_j,e_j)\mathbb{Z}F+\mathfrak{s}+\mathfrak{f}^n)$ then $[x_i,x_j]^{e_i d_{ij}} \in S'\,\gamma_{n+2}(F)$. Indeed, let $d_{ij} = t(x_i,e_i)a_{ij}$, $a_{ij} \in \mathbb{Z}F$. Then, modulo $\mathfrak{s}$, since $x_i^k t(x_i,e_i) \equiv t(x_i,e_i)$, we have

$$\begin{aligned} e_i d_{ij} &\equiv e_i t(x_i,e_i)a_{ij} \\ &\equiv t(x_i,e_i)t(x_i,e_i)a_{ij} \\ &\equiv t(x_i,e_i)d_{ij}. \end{aligned}$$

Thus,

$$e_i d_{ij} \equiv t(x_i,e_i)t(x_j,e_j)b_{ij} \pmod{\mathfrak{s}+\mathfrak{f}^n}$$

for some $b_{ij} \in \mathbb{Z}F$ and, in turn,

$$\begin{aligned} [x_i,x_j]^{e_i d_{ij}} &\equiv [x_i,x_j]^{t(x_i,e_i)t(x_j,e_j)b_{ij}} \quad \mathrm{mod}[F',S]\gamma_{n+2}(F) \\ &\equiv [x_i^{e_i},x_j^{e_j}]^{b_{ij}} \ \mathrm{mod}[F',S]\gamma_{n+2}(F) \\ &\equiv 1 \ \mathrm{mod}\ S'\,\gamma_{n+2}(F). \end{aligned}$$

□

Since $S' \le R$, in view of Proposition 3.1, Proposition 3.4 yields,

3.5 *COROLLARY*. (Gupta 1984). If G is a metabelian group with G/G' of exponent e then $D_{n+2}(G)/\gamma_{n+2}(G)$ has exponent dividing e for all $n \ge 1$. □

3.6 *THEOREM*. (Gupta 1984). $[D(n+2,\mathfrak{f}\mathfrak{s}),F] \le [F',S]\gamma_{n+3}(F)$ for all $n \ge 1$.

Proof. It clearly suffices to show that for each w_i of the form (16), $[w_i,x_k] \in [F',S]\gamma_{n+3}(F)$ for all $k = 1,\ldots,m$. Indeed, we have, modulo F",

$$\begin{aligned}[w_i,x_k] &\equiv \prod_{j=i+1}^{m} [[x_i,x_j]^{d_{ij}},x_k] \\ &\equiv \prod_{j=i+1}^{m} [x_i,x_j,x_k]^{d_{ij}} \\ &\equiv \prod_{j=i+1}^{m} [x_i,x_k,x_j]^{d_{ij}}[x_k,x_j,x_i]^{d_{ij}} \\ &\equiv \prod_{j=i+1}^{m} [x_i,x_k]^{(x_j-1)d_{ij}}[x_k,x_j]^{(x_i-1)d_{ij}}.\end{aligned}$$

Since, by Theorem 3.2, $d_{ij} \in t(x_i,e_i)\mathbb{Z}F \cap (t(x_j,e_j)\mathbb{Z}F + \underline{\underline{s}} + \underline{\underline{f}}^n)$, it follows that $(x_j-1)d_{ij}$, $(x_i-1)d_{ij} \in \underline{\underline{s}} + \underline{\underline{f}}^{n+1}$ and consequently $[w_i,x_k] \in [F',S]\gamma_{n+3}(F)$. □

An immediate consequence of Theorem 3.6 is the following result of independent interest due to Gupta & Passi.

3.7 *COROLLARY*. If G is a metabelian group then $[D_n(G),G] = \gamma_{n+1}(G)$ for all n. □

We next prove a result about cyclic group rings.

Let $\mathbb{Z}G$ be the group ring of an infinite cyclic group $G = \langle x\rangle$. For each $n \geq 1$, define

$$C(n)^* = b(1)\ldots b(n), \quad b(k) = \ell.c.m.\{1,\ldots,k\}. \tag{19}$$

Then we prove the following result of independent interest.

3.8 *PROPOSITION*. (Gupta 1984). For all $n \geq 1$, $e \geq 1$,

$$C(n)^*t(x,e) \equiv C(n)^*e \,(\text{modulo } (x^e-1)\mathbb{Z}G + (x-1)^n\mathbb{Z}G),$$

where $t(x,e) = 1 + x + \ldots + x^{e-1}$.

Proof. For each $k \geq 1$, let $\binom{e}{k}^{\#}$ be the additive subgroup of $\mathbb{Z}$ defined by

$$\binom{e}{k}^{\#} = \left\langle\binom{e}{1},\ldots,\binom{e}{k}\right\rangle \tag{20}$$

($\binom{e}{i} = 0$ for $i > e$); and for each $n \geq 1$, let J_n be the additive subgroup of $\mathbb{Z}G$ defined by

$$J_n = \langle \binom{e}{n}^{\#}(x-1)^n, \binom{e}{n+1}^{\#}(x-1)^{n+1}, \ldots \rangle . \tag{21}$$

Then clearly

$$\binom{e}{1}^{\#} \subseteq \binom{e}{2}^{\#} \subseteq \cdots \tag{22}$$

and

$$(x-1)^j J_n \subseteq J_{n+j} \tag{23}$$

for all $j \geq 1$, $n \geq 1$.

The expansion

$$(x^e-1) = e(x-1) + \sum_{k=2}^{\infty} \binom{e}{k}(x-1)^k$$

yields

$$e(x-1) \equiv (x^e-1) \bmod J_2, \tag{24}$$

which on pre-multiplication with $(x-1)^j$ for any $j \geq 1$, gives using (20), (23),

$$e(x-1)^{j+1} \equiv 0 \bmod ((x-1)(x^e-1)\mathbb{Z}G + J_{j+2}). \tag{25}$$

Define $b(k) = \ell.c.m.\{1,\ldots,k\}$. Then using the fact that $\binom{e}{j} = \frac{e}{j}\binom{e-1}{j-1}$ for all j, we see that

$$b(k)\binom{e}{k}^{\#} \subseteq \binom{e}{1}^{\#} = e\mathbb{Z} . \tag{26}$$

We prove by induction on $n \geq 1$ that

$$C(n)^* e(x-1) \equiv C(n)^*(x^e-1) \bmod ((x-1)(x^e-1)\mathbb{Z}G + J_{n+1}), \tag{27}$$

where $C(n)^*$ is defined as in (19).

For $n = 1$, (27) follows from (24). For the inductive step, we assume (27) for some $n \geq 1$. Then

$$b(n+1)C(n)^* e(x-1) \equiv b(n+1)C(n)^*(x^e-1)$$

modulo $(x-1)(x^e-1)\mathbb{Z}G + b(n+1)J_{n+1}$.

Thus it suffices to prove that

$$b(n+1)J_{n+1} \subseteq (x-1)(x^e-1)\mathbb{Z}G + J_{n+2}.$$

Indeed,

$$b(n+1)J_{n+1} \subseteq b(n+1)\binom{e}{n+1}^{\#}(x-1)^{n+1} + J_{n+2}$$

$$\subseteq \binom{e}{1}^{\#}(x-1)^{n+1} + J_{n+2}, \quad \text{by (26)}$$

$$\subseteq (x-1)(x^e-1)\mathbb{Z}G + J_{n+2}, \quad \text{by (25)}.$$

This completes the proof of (27). Dividing both sides of (27) by $(x-1)$ yields

$$C(n)^* e \equiv C(n)^*(1+x+\ldots+x^{e-1}) \bmod ((x^e-1)\mathbb{Z}G + (x-1)^n\mathbb{Z}G),$$

as was to be proved. □

3.9 *REMARKS*. A.W. Hales (verbal communication) has shown that the least positive integer $C(n)^*$ satisfying the assertion of Proposition 3.8 is $\prod_{p<n} p^{[n-1/p-1]}$. □

4. DIMENSION SUBGROUPS OF METABELIAN GROUPS

Let G be a metabelian group given by a presentation of the form (10). Since $S' \leq R$, by Proposition 3.1, it follows that if for some $\ell \geq 1$, $D(n+2,\mathfrak{f}\mathfrak{s})^{\ell} \leq S'\gamma_{n+2}(F)$, $\mathfrak{s} = \mathbb{Z}F(S-1)$, then $D_{n+2}(G)^{\ell} \leq \gamma_{n+2}(G)$. This observation suggests the study of the quotients $D(n+2,\mathfrak{f}\mathfrak{s})/S'\gamma_{n+2}(F)$ as an aid to the study of the quotients $D_{n+2}(G)/\gamma_{n+2}(G)$. We start with the following theorem.

4.1 *LEMMA*. (Gupta, Hales & Passi 1984). Let $S = \langle x_1^{e_1},\ldots,x_m^{e_m},F'\rangle \leq F = \langle x_1,\ldots,x_m\rangle$, where $e_m|\ldots|e_1 \geq 0$. Then there exists an integer $n_0 = n_0(e_1,\ldots,e_m)$ such that, for all $n \geq n_0$, $D(n+2,\mathfrak{f}\mathfrak{s}) = S'\gamma_{n+2}(F)$.

Proof. We note that the ring $\mathbb{Z}F/\mathfrak{s}$ $(\cong \mathbb{Z}(F/S))$ is a commutative Noetherian ring. Let $U_i = t(x_i,e_i)\mathbb{Z}F+\mathfrak{s}/\mathfrak{s}$ and $V = \mathfrak{f}/\mathfrak{s}$ be ideals of $\mathbb{Z}F/\mathfrak{s}$. By the Artin-Rees lemma, there exists n_0 such that

$$U_i \cap V^n = (U_i \cap V^{n_0})V^{n-n_0}$$

for all $n \geq n_0$. (See, for instance, Lang (1984), page 241.)

Let $a_{ij} \in \mathbb{Z}F$ be such that

$$t(x_i,e_i)a_{ij} \in t(x_j,e_j)\mathbb{Z}F + \mathfrak{s} + \mathfrak{f}^n \tag{28}$$

for some $j > i$.

Then $(x_j-1)t(x_i,e_i)a_{ij} \in \underline{\underline{s}} + \underline{\underline{f}}^{n+1}$ and it follows that $(x_j-1)t(x_i,e_i)a_{ij} + \underline{\underline{s}}/\underline{\underline{s}} \in U_i \cap V^{n+1} = (U_i \cap V^{n_0})V^{n+1-n_0}$ for all $n+1 \geq n_0$. Consequently, modulo $\underline{\underline{s}}$, we have for all $n+1 \geq n_0$,

$$(x_j-1)t(x_i,e_i)a_{ij} \equiv \sum_k t(x_i,e_i)\alpha_k\beta_k \tag{29}$$

with $t(x_i,e_i)\alpha_k \in \underline{\underline{f}}^{n_0} + \underline{\underline{s}}$ and $\beta_k \in \underline{\underline{f}}^{n+1-n_0}$.

Assuming $n+1 \geq n_0$, we can write

$$\beta_k = \sum_{\ell=1}^{m} (x_\ell-1)\gamma_{k\ell}, \quad \gamma_{k\ell} \in \underline{\underline{f}}^{n-n_0}. \tag{30}$$

Now, modulo $\underline{\underline{f}}\underline{\underline{s}}$, we have

$$[x_i,x_j]^{t(x_i,e_i)a_{ij}} - 1$$

$$\equiv ((x_i-1)(x_j-1) - (x_j-1)(x_i-1))t(x_i,e_i)a_{ij}$$

$$\equiv (x_i-1)(x_j-1)t(x_i,e_i)a_{ij}$$

$$\equiv (x_i-1)\sum_k t(x_i,e_i)\alpha_k\beta_k, \quad \text{by (29)}$$

$$\equiv \sum_{k,\ell} \{(x_i-1)(x_\ell-1) - (x_\ell-1)(x_i-1)\}t(x_i,e_i)\alpha_k\gamma_{k\ell}, \quad \text{by (30)}$$

$$\equiv \prod_\ell [x_i,x_\ell]^{\sum_k t(x_i,e_i)\alpha_k\gamma_{k\ell}} - 1$$

$$\equiv f_{ij} - 1,$$

where $f_{ij} \in [F',S]\gamma_{n+2}(F)$, since $t(x_i,e_i)\alpha_k\gamma_{k\ell} \in \underline{\underline{f}}^n + \underline{\underline{s}}$. Thus

$$[x_i,x_j]^{t(x_i,e_i)a_{ij}}f_{ij}^{-1} \in F \cap (1+\underline{\underline{f}}\underline{\underline{s}}) = S'$$

(by I.1.9), and it follows that

$$[x_i,x_j]^{t(x_i,e_i)a_{ij}} \in S'\,\gamma_{n+2}(F),$$

for all $a_{ij} \in \mathbb{Z}F$ satisfying (28).

By Theorem 3.2, $D(n+2,\underline{\underline{f}}\underline{\underline{s}})$ is generated by the elements $[x_i,x_j]^{t(x_i,e_i)a_{ij}}$, $1 \leq i < j \leq m$, with a_{ij} satisfying (28) and it follows that $D(n+2,\underline{\underline{f}}\underline{\underline{s}}) = S'\,\gamma_{n+2}(F)$ for all $n \geq n_0$. □

4.2 *REMARKS*. Hales (1985) has shown that if $S = \langle x_1^{p^{e_1}}, \ldots, x_m^{p^{e_m}}, F' \rangle$, $e_m \le \ldots \le e_1$, p prime, then the best possible choice for n_0 in Theorem 4.1 is $n_0 = p^{e_1} + p^{e_1 - 1}$. □

As an immediate Corollary to Theorem 4.1 we have

4.3 *THEOREM*. (Gupta, Hales & Passi 1984). If G is a finitely generated metabelian group then there exists an integer $n_0 = n_0(G/G')$ such that $D_n(G) = \gamma_n(G)$ for all $n \ge n_0$. □

Next, we turn to the study of $D(n+2, \underline{\underline{f}}\underline{\underline{s}})/D(n+2, \underline{\underline{f}}^2\underline{\underline{s}})$ which is required for improving Sjogren's bound for the metabelian case. Let $w \in D(n+2, \underline{\underline{f}}\underline{\underline{s}})$. Then by Theorem 3.2 we may write w as

$$w = \prod_{1 \le i < j \le m} [x_i, x_j]^{d_{ij}} \prod_{1 \le k \le m} [x_k^{e_k}, \eta_k]\, \xi, \tag{31}$$

where $d_{ij} = t(x_i, e_i) a_{ij}$, $a_{ij} = a_{ij}(x_j, \ldots, x_m) \in \mathbb{Z}F$; $d_{ij} \equiv t(x_j, e_j) b_{ij} \mod(\underline{\underline{s}} + \underline{\underline{f}}^n)$, $b_{ij} \in \mathbb{Z}F$, $\eta_k \in F'$ and $\xi \in F''\gamma_{n+2}(F)$.

Modulo $\underline{\underline{f}}^2\underline{\underline{a}}$, $\underline{\underline{a}} = \mathbb{Z}F(F'-1)$, we have as in Gupta & Tahara (1985),

$$\begin{aligned}
&[x_i, x_j]^{x_1^{\alpha_1} \ldots x_m^{\alpha_m}} - 1 \\
&\equiv x_m^{-\alpha_m} \ldots x_1^{-\alpha_1}([x_i, x_j]-1) x_1^{\alpha_1} \ldots x_m^{\alpha_m} \\
&\equiv ([x_i, x_j]-1) x_1^{\alpha_1} \ldots x_m^{\alpha_m} - \sum_{k=1}^{m} (x_k - 1)([x_i, x_j]-1) \times \alpha_k x_1^{\alpha_1} \ldots x_m^{\alpha_m} \\
&\equiv ([x_i, x_j]-1) x_1^{\alpha_1} \ldots x_m^{\alpha_m} \\
&\quad - \sum_{k=1}^{m} (x_k - 1)([x_i, x_j]^{x_k D_{x_k}(x_1^{\alpha_1} \ldots x_m^{\alpha_m})} - 1)
\end{aligned}$$

where $D_{x_k}(u)$ is the usual partial derivative as defined for rational functions. Since $d_{ij} \in (t(x_i, e_i)\mathbb{Z}F \cap (t(x_j, e_j)\mathbb{Z}F + \underline{\underline{s}} + \underline{\underline{f}}^n)$, it follows that, modulo $\underline{\underline{f}}^2\underline{\underline{s}} + \underline{\underline{f}}^{n+2}$,

$$\begin{aligned}
&[x_i, x_j]^{d_{ij}} - 1 \\
&\equiv ([x_i, x_j]-1) d_{ij} - \sum_{k=1}^{m} (x_k - 1)([x_i, x_j]^{x_k D_{x_k}(d_{ij})} - 1)
\end{aligned}$$

$$\equiv ([x_i,x_j]-1)d_{ij} - \sum_{k=1}^{m} (x_k-1)([x_i,x_j]^{x_k D_{x_k}(d_{ij})} - 1)$$

$$\equiv ((x_i-1)(x_j-1) - (x_j-1)(x_i-1))d_{ij}$$

$$- \sum_{k=i}^{m} (x_k-1)([x_i,x_j]^{x_k D_{x_k}(d_{ij})} - 1),$$

since $d_{ij} = d_{ij}(x_i,\ldots,x_m)$ and $D_{x_k}(d_{ij}) = 0$ for $k < i$.

Also, since $\eta_k - 1 \in \underline{\underline{\mathfrak{f}}}^2$, modulo $\underline{\underline{\mathfrak{f}}}^2\underline{\underline{\mathfrak{s}}}$,

$$[x_k^{e_k},\eta_k] - 1 \equiv (x_k^{e_k}-1)(\eta_k-1)$$

$$\equiv (x_k-1)(\eta_k^{e_k}-1).$$

Thus if w is of the form (31) then modulo $\underline{\underline{\mathfrak{f}}}^2\underline{\underline{\mathfrak{s}}}+\underline{\underline{\mathfrak{f}}}^{n+2}$,

$$w-1 \equiv \sum_{1\le i<j\le m} ((x_i-1)(x_j-1) - (x_j-1)(x_i-1))d_{ij}$$

$$- \sum_{1\le i<j\le m} \sum_{k=i}^{m} (x_k-1)([x_i,x_j]^{x_k D_{x_k}(d_{ij})} - 1)$$

$$+ \sum_{1\le k\le m} (x_k-1)(\eta_k^{e_k}-1),$$

$$\equiv \sum_{1\le i<j\le m} (x_i-1)(x_j^{e_j b_{ij}}-1) - (x_j-1)(x_i^{e_i a_{ij}}-1)$$

$$- \sum_{1\le i<j\le m} \sum_{k=i}^{m} (x_k-1)([x_i,x_j]^{x_k D_{x_k}(d_{ij})} - 1)$$

$$+ \sum_{1\le k\le m} (x_k-1)(\eta_k^{e_k}-1)$$

$$\equiv \sum_{k=1}^{m} (x_k-1)(y_k z_k^{-1} \eta_k^{e_k}-1),$$

where

$$y_k = \prod_{i<k} x_i^{-e_i a_{ik}} \prod_{k<j} x_j^{e_j b_{kj}}; \tag{32}$$

$$z_k = \prod_{\substack{i<j \\ i\le k}} [x_i,x_j]^{x_k D_{x_k}(d_{ij})}. \tag{33}$$

We record this as

4.4 <u>*LEMMA*</u>. Let $w \in D(n+2,\mathfrak{f}\mathfrak{s})$ be as in (31). Then, modulo $\mathfrak{f}^2\mathfrak{s}+\mathfrak{f}^{n+2}$, $w-1 \equiv \sum\limits_{k=1}^{m} (x_k-1)(y_k z_k^{-1}\eta_k^{e_k}-1)$, where y_k, z_k are as in (32), (33) and $\eta_k \in F'$. □

The elements y_k are related to w by the following lemma.

4.5 <u>*LEMMA*</u>. $w^2 \equiv \sum\limits_{k=1}^{m} [x_k, y_k] \pmod{[F',S]\gamma_{n+2}(F)}$.

<u>*Proof*</u>. We have, modulo $[F',S]\gamma_{n+2}(F)$,

$$\prod_{k=1}^{m} [x_k,y_k] \equiv \prod_{k=1}^{m}\left(\prod_{i<k} [x_k, x_i^{-e_i a_{ik}}] \prod_{k<j} [x_k, x_j^{e_j b_{kj}}]\right)$$

$$\equiv \prod_{1\le i<k\le m} [x_i,x_k]^{t(x_i,e_i)a_{ik}} \prod_{1\le k<j\le m} [x_k,x_j]^{t(x_j,e_j)b_{kj}}$$

$$\equiv \prod_{1\le i<k\le m} [x_i,x_k]^{d_{ik}} \prod_{1\le k<j\le m} [x_k,x_j]^{d_{kj}}$$

$$\equiv w^2. \qquad \square$$

In Section 1 we proved Sjogren's theorem that $D_{n+2}(G)/\gamma_{n+2}(G)$ has exponent dividing $C(n) = b(1)^{\binom{n}{1}} \ldots b(n)^{\binom{n}{n}}$, where $b(k) = \ell.c.m.\{1,\ldots,k\}$. If G is assumed to be metabelian then we can improve Sjogren's bounds as follows.

4.6 <u>*THEOREM*</u>. (Gupta 1984). If G is a metabelian group then for all $n \ge 1$, $D_{n+2}(G)/\gamma_{n+2}(G)$ has exponent dividing $2C(n)^*$, where $C(n)^* = b(1)\ldots b(n)$.

<u>*Proof*</u>. Let $G = F/R$ be given by a presentation of the form (10). Then it suffices to prove that if $w \in D(n+2,\mathfrak{f}\mathfrak{r})$ then $w^{2C(n)^*} \in R\,\gamma_{n+2}(F)$. We proceed as follows. We have

$$D(n+2,\mathfrak{f}\mathfrak{r}) \le D(n+2,\mathfrak{f}\mathfrak{s}) \cap D(n+2,\mathfrak{f}\mathfrak{r}+\mathfrak{f}^2\mathfrak{s}).$$

Thus by Theorem 3.2

$$w \equiv \prod_{1\le i<j\le m} [x_i,x_j]^{d_{ij}} \prod_{1\le k\le m} [x_k^{e_k},\eta_k] \ (\text{modulo } F'' \ \gamma_{n+2}(F))$$

where

$$d_{ij} = t(x_i,e_i)a_{ij}, \quad a_{ij} = a_{ij}(x_j,\ldots,x_m) \in \mathbb{Z}F\,;$$

$$d_{ij} \equiv t(x_j,e_j)b_{ij} \ (\text{mod } \underline{\underline{s}}+\underline{\underline{f}}^n), \ b_{ij} \in \mathbb{Z}F\,;$$

$$\eta_k \in F' .$$

By Lemma 4.4,

$$w-1 \equiv \sum_{k=1}^{m} (x_k-1)(y_k z_k^{-1}\eta_k^{e_k}-1)\,(\text{mod } \underline{\underline{f}}^2\underline{\underline{s}} + \underline{\underline{f}}^{n+2}),$$

where y_k, z_k are as in (32), (33).

Thus

$$\sum_{k=1}^{m} (x_k-1)(y_k z_k^{-1}\eta_k^{e_k}-1) \equiv 0 \text{ mod } \underline{\underline{f}}(\underline{\underline{r}}+\underline{\underline{f}}\underline{\underline{s}}+\underline{\underline{f}}^{n+1}),$$

and, in turn,

$$y_k z_k^{-1}\eta_k^{e_k} - 1 \in \underline{\underline{r}} + \underline{\underline{f}}\underline{\underline{s}} + \underline{\underline{f}}^{n+2},$$

for all $k = 1,\ldots,m$; and consequently

$$y_k z_k^{-1}\eta_k^{e_k} r_k - 1 \in \underline{\underline{f}}\underline{\underline{s}} + \underline{\underline{f}}^{n+1} \tag{34}$$

for some $r_k \in R$.

By Theorem 3.6, it follows using $S' \le R$ that

$$[x_k, y_k z_k^{-1}\eta_k^{e_k} r_k] \in R\,\gamma_{n+2}(F)$$

and since $[x_k,r_k] \in R$, we have

$$[x_k, y_k z_k^{-1}\eta_k^{e_k}] \in R\,\gamma_{n+2}(F). \tag{35}$$

Now

$$\begin{aligned}[x_k,\eta_k^{e_k}]^{C(n-1)^*} &\equiv [x_k,\eta_k]^{e_k C(n-1)^*} \ (\text{mod } F'') \\ &\equiv [x_k,\eta_k]^{t(x_k,e_k)C(n-1)^*} \\ &\qquad (\text{mod} [F',S]\gamma_{n+2}(F)), \quad (\text{Proposition 3.8})\end{aligned}$$

$$\equiv [x_k^{e_k}, \eta_k]^{C(n-1)^*} \bmod [F',S]\gamma_{n+2}(F),$$

by (12),

$$\equiv 1 \pmod{[F',S]\gamma_{n+2}(F)},$$

and it follows that

$$[x_k, \eta_k^{e_k}]^{C(n-1)^*} \in R\,\gamma_{n+2}(F). \tag{36}$$

Further, for $i < j$, $i \le k$, we have

$$b(n)D_{x_k}(d_{ij}) = b(n)D_{x_k}(t(x_i,e_i)a_{ij})$$

$$= \begin{cases} b(n)t(x_i,e_i)D_{x_k}(a_{ij}), & i < k \\ b(n)D_{x_k}(t(x_k,e_k))a_{kj}, & i = k \end{cases}$$

$$= \begin{cases} b(n)\displaystyle\sum_{\ell=1}^{\infty}\binom{e_i}{\ell}(x_i-1)^{\ell-1}D_{x_k}(a_{ij}), & i < k \\ b(n)\displaystyle\sum_{\ell=2}^{\infty}(\ell-1)\binom{e_k}{\ell}(x_k-1)^{\ell-2}a_{kj}, & i = k \end{cases}$$

$$\in e_k\mathbb{Z}F + \underline{\mathfrak{f}}^{n-1}, \qquad \text{since } e_k | e_i.$$

Thus, modulo $[F',S]\gamma_{n+2}(F)$,

$$[x_k,z_k]^{C(n)^*} \equiv \prod_{i\le k}[x_k,[x_i,x_j]]^{b(n)D_{x_k}(d_{ij})x_kC(n-1)^*}$$

$$\equiv \prod_{i\le k}[x_k,[x_i,x_j]]^{e_kC(n-1)^*u_{ijk}}, \quad u_{ijk} \in \mathbb{Z}F,$$

$$\equiv \prod_{i\le k}[x_k,[x_i,x_j]]^{t(x_k,e_k)C(n-1)^*u_{ijk}},$$

by Proposition 3.8,

$$\equiv \prod_{i\le k}[x_k^{e_k},[x_i,x_j]]^{C(n-1)^*u_{ijk}}$$

$$\equiv 1.$$

It follows that

$$[x,z_k^{-1}]^{C(n)^*} \in R\,\gamma_{n+2}(F). \tag{37}$$

Using (36) and (37), (35) yields

$$[x_k, y_k]^{C(n)^*} \in R\,\gamma_{n+2}(F) \tag{38}$$

for all $k = 1,\ldots,m$. Thus by Lemma 4.5 we have

$$w^{2C(n)^*} \equiv \prod_{k=1}^{m} [x_k, y_k]^{C(n)^*} \equiv 1 \bmod R\,\gamma_{n+2}(F). \qquad \square$$

The conclusion (35) together with Lemma 4.5 yields the following criterion.

4.7 *COROLLARY*. Let $w \in D(n+2,\mathfrak{r})$. Then $w^2 \in R\,\gamma_{n+2}(F)$ if and only if $\prod_{k=1}^{m} [x_k, z_k\eta_k^{-e_k}] \in R\,\gamma_{n+2}(F)$ where z_k is as in (33) and η_k is as in (31). $\square$

4.8 *REMARKS*. Since, modulo $R\,\gamma_4(F)$, $[x_k, \eta_k^{e_k}] \equiv [x_k^{e_k}, \eta_k] \equiv 1$ and $[x_k, [x_i, x_j]^{D_{x_k}(d_{ij})}] \equiv 1$, for $i \le k$, Corollary 4.7 yields an alternate proof of Losey's result (Corollary 1.11). $\square$

5. STRUCTURE OF THE FOURTH DIMENSION SUBGROUP

A complete description of $D_4(G)/\gamma_4(G)$ for finite G has been given by Tahara (1977). If G is nilpotent of class 3 then G is necessarily metabelian and consequently the theory developed in the preceeding section applies to G. In what follows we shall obtain an alternate description of $D_4(G)/\gamma_4(G)$.

5.1 *THEOREM*. (Tahara 1977, Gupta 1984). Let $G = F/R$ with $F/RF' \cong F/S$, $F = \langle x_1,\ldots,x_m\rangle$, $S = \langle x_1^{e_1},\ldots,x_m^{e_m}, F'\rangle$, $e_i = 2^{\alpha_i}$, $\alpha_1 \ge \alpha_2 \ge \cdots \ge \alpha_m \ge 1$. Then modulo $R\,\gamma_4(F)$, $D(4,R)$ consists of all elements $w = \prod_{1\le i<j\le m} [x_i^{e_i}, x_j]^{a_{ij}}$, $a_{ij} \in \mathbb{Z}$, such that

(i) $e_j \mid \binom{e_i}{2} a_{ij}$, $1 \le i < j \le m$;

(ii) $y_k = \prod_{i<k} x_i^{-e_i a_{ik}} \prod_{k<j} x_j^{e_j b_{kj}} \in R\,F'^{e_k}\,\gamma_3(F)$, $1 \le k \le m$,

where $b_{kj} = (e_k/e_j)a_{kj} + (\binom{e_k}{2}/e_j)a_{kj}(x_k - 1)$.

Proof. Let $w \in D(4,\mathfrak{r}) \le D(4,\mathfrak{s})$. Then, by Theorem 3.2,

$$w = \prod_{1\le i<j\le m} [x_i, x_j]^{d_{ij}} \prod_{k=1}^{m} [x_k^{e_k}, \eta_k]\xi,$$

where $\xi \in \gamma_4(F)$, $\eta_k \in F'$, $d_{ij} = t(x_i,e_i)a_{ij} \equiv t(x_j,e_j)b_{ij}\,(\underline{\underline{s}}+\underline{\underline{f}}^2)$, $a_{ij} = a_{ij}(x_j,\ldots,x_m) \in \mathbb{Z}F$, $b_{ij} = b_{ij}(x_i,\ldots,x_m) \in \mathbb{Z}F$.

Since

$$[x_i,x_j]^{t(x_i,e_i)(x_k-1)} \equiv [x_i,x_j]^{e_i(x_k-1)} \bmod \gamma_4(F)$$

$$\equiv [x_i,x_j]^{(x_k^{e_k}-1)(e_i/e_k)} \bmod \gamma_4(F)$$

$$\equiv 1 \bmod [F',S]\gamma_4(F),$$

it follows that $a_{ij} \in \mathbb{Z}$ and $b_{ij} = b'_{ij}+b''_{ij}(x_i-1)$ with $b'_{ij}, b''_{ij} \in \mathbb{Z}$. Equating $[x_i,x_j]^{t(x_i,e_i)a_{ij}}$ and $[x_i,x_j]^{t(x_j,e_j)b_{ij}}$ yields

$$b_{ij} = (e_i/e_j)a_{ij} + (\tbinom{e_i}{2}/e_j)a_{ij}(x_i-1)$$

and

$$e_j \mid \tbinom{e_j}{2}(e_i/e_j)a_{ij}$$

so (i) is satisfied and b_{ij} is as in (ii).

As in (34) (see proof of Theorem 4.6), we have

$$y_k z_k^{-1}\eta_k^{e_k} r_k - 1 \in \underline{\underline{f}}\,\underline{\underline{s}} + \underline{\underline{f}}^3 \leq \underline{\underline{r}} + \underline{\underline{f}}^3,$$

so

$$y_k z_k^{-1}\eta_k^{e_k} r_k \in R\,\gamma_3(F).$$

Or, equivalently,

$$y_k z_k^{-1} \in R\,F'^{e_k}\,\gamma_3(F).$$

Now,

$$z_k = \prod_{\substack{i<j \\ i\leq k}} [x_i,x_j]^{x_k D_{x_k}(d_{ij})}$$

and, modulo $R\,\gamma_3(F)$,

$$[x_i,x_j]^{D_{x_k}(d_{ij})} = \begin{cases} [x_i,x_j]^{t(x_i,e_i)D_{x_k}(a_{ij})}, & k \neq i \\ [x_i,x_j]^{t(x_j,e_j)D_{x_k}(b_{ij})}, & k = i \end{cases}$$

$$\equiv \quad 1.$$

Thus $y_k \in R\,F'^{e_k}\,\gamma_3(F)$ as required.

Conversely, let $w = \prod_{1\le i<j\le m} [x_i^{e_i},x_j]^{a_{ij}}$, $a_{ij} \in \mathbb{Z}$, satisfying (i) and (ii). It is readily verified that

$$t(x_i,e_i)a_{ij} \equiv t(x_j,e_j)b_{ij} \pmod{\underline{\underline{s}}+\underline{\underline{f}}^2}.$$

Thus, by Theorem 3.2, $w \in D(4,\underline{\underline{f}}\underline{\underline{s}})$ and consequently, by Lemma 4.4,

$$w-1 \equiv \sum_{k=1}^{m} (x_k-1)(y_k z_k^{-1} n_k^{e_k}-1) \bmod \underline{\underline{f}}^2\underline{\underline{s}} + \underline{\underline{f}}^4,$$

where z_k is of the form (33) and lies in $R\,\gamma_3(F)$ as before so that $(x_k-1)(z_k^{-1}-1) \in \underline{\underline{f}}\underline{\underline{r}}+\underline{\underline{f}}^4$. By hypothesis, $(x_k-1)(y_k-1) \in \underline{\underline{f}}\underline{\underline{r}}+\underline{\underline{f}}^4$. Finally, $(x_k-1)(n_k^{e_k}-1) \equiv (x_k^{e_k}-1)(n_k-1) \equiv 0 \bmod \underline{\underline{r}}+\underline{\underline{f}}^4$. Since $\underline{\underline{f}}^2\underline{\underline{s}}+\underline{\underline{f}}^4 \le \underline{\underline{f}}^2\underline{\underline{r}}+\underline{\underline{f}}^4 \le \underline{\underline{r}}+\underline{\underline{f}}^4$, we conclude that $w-1 \in \underline{\underline{r}}+\underline{\underline{f}}^4$, i.e., $w \in D(4,R)$. □

Rips' counter-example constructed in Section 2 is a 4 generator group. We now show that if G is 2 or 3 generator group then $D_4(G) = \gamma_4(G)$.

5.2 *COROLLARY*. (Tahara 1977). If in Theorem 5.1, $m \le 3$, then $D(4,R) = R\,\gamma_4(F)$.

Proof. (Passi). If $m = 2$ then $F'^{e_2} \le R\,\gamma_3(F)$ and the condition (ii) implies $y_2 = x_1^{-e_1 a_{12}} \in R\,\gamma_3(F)$. Hence

$$w = [x_1^{e_1},x_1]^{a_{12}} \equiv [x_1^{e_1 a_{12}},x_2] \equiv 1 \pmod{R\,\gamma_4(F)}.$$

Let $m = 3$ and consider the element

$$w = [x_1^{e_1},x_2]^{a_{12}}[x_1^{e_1},x_3]^{a_{13}}[x_2^{e_2},x_3]^{a_{23}}$$

and a_{ij} is even if $e_i = e_j$. Suppose

$$y_1 = x_2^{e_1a_{12}} x_3^{e_1a_{13}} \in R\, F'^{e_1}\, \gamma_3(F);$$

$$y_2 = x_1^{-e_1a_{12}} x_3^{e_2a_{23}} \in R\, F'^{e_2}\, \gamma_3(F)$$

$$y_3 = x_1^{-e_1a_{13}} x_2^{-e_2a_{23}} \in R\, F'^{e_3}\, \gamma_3(F).$$

Assume $w \notin R\,\gamma_4(F)$. Then, since $[x_1,y_1] \in R\,\gamma_4(F)$ (using $[x_1,\eta^{e_1}] \equiv [x_1^{e_1},\eta] \equiv 1 \bmod R\,\gamma_4(F))$, we obtain

$$[x_1,x_2]^{e_1a_{12}}[x_1,x_3]^{e_1a_{13}} \equiv 1 \pmod{R\,\gamma_4(F)}$$

and since $w \notin R\,\gamma_4(F)$, it follows that

$$[x_2,x_3]^{e_2a_{23}} \equiv 1 \pmod{R\,\gamma_4(F)}.$$

On the other hand, $[x_2,y_1] \in R\,\gamma_4(F)$ implies

$$[x_2,x_3]^{e_1a_{13}} \equiv 1 \pmod{R\,\gamma_4(F)}.$$

It follows that

$$v_2(e_1a_{13}) > v_2(e_2a_{23}), \tag{39}$$

where $v_2(n)$ is the highest power of 2 dividing n.

Similarly, using $[x_2,y_2] \in R\,\gamma_4(F)$ gives

$$[x_1,x_3]^{e_1a_{13}} \not\equiv 1 \pmod{R\,\gamma_4(F)}$$

and $[x_1,y_2] \in R\,\gamma_4(F)$ gives

$$[x_1,x_3]^{e_2a_{23}} \equiv 1 \pmod{R\,\gamma_4(F)}$$

and consequently

$$v_2(e_2a_{23}) > v_2(e_1a_{13}). \tag{40}$$

Since (39) and (40) are contradictory we must have $w \in R\,\gamma_4(F)$.□

5.3 *REMARKS*. Another proof of Corollary 5.2 is given in Passi, Sucheta & Tahara (1986). For $m = 2$, the first proof of Corollary 5.2 was given by Passi (1968).

6. DIMENSION SUBGROUPS OF METABELIAN p-GROUPS

In this section we prove that the n-th dimension subgroup of a finite metabelian p-group coincides with its n-th lower central subgroup for all $n \le 2p-1$. This is our most recent result in the theory of dimension subgroups and its proof uses a careful blend of techniques developed in Sections 1 and 3.

Let G be a finite non-cyclic metabelian group given by a pre-abelian free presentation

$$1 \to RF'' \to F \to G \to 1$$

with $F = \langle x_1,\dots,x_m\rangle$, $m \ge 2$, F/R finite and R is the normal closure

$$R = \langle x_1^{e(1)}\xi_1,\dots,x_m^{e(m)}\xi_m,\xi_{m+1},\dots,\xi_s\rangle^F \tag{41}$$

satisfying $e(m)|\dots|e(2)|e(1) > 0$, $\xi_i \in F'$.

Since F/R is finite, we have for each i, $x_i^{\bar{e}(i)} \in R$ for some $\bar{e}(i) > 0$. Thus, replacing x_i^{-1} by $x_i^{\bar{e}(i)-1}$ in (41) we may express R as

$$R = \langle x_1^{e(1)}\eta_1,\dots,x_m^{e(m)}\eta_m,\eta_{m+1},\dots,\eta_s,x_1^{\bar{e}(1)},\dots,x_m^{\bar{e}(m)}\rangle^F$$

where each η_i is a positive word in $x_1,\dots,x_m$. In other words, R is the normal closure of the form

$$R = \langle r_1,\dots,r_{s+m}\rangle^F \tag{42}$$

where $r_i = x_{i(1)}\cdots x_{i(\ell_i)}$ and $x_{i(j)} \in \{x_1,\dots,x_m\}$. We shall need to use each of the formats (41) and (42) of R.

As in Section 1, define

$$\underline{\mathfrak{r}}(1) = \underline{\mathfrak{r}} = \mathbb{Z}F(R-1)$$

and, for $k \ge 2$,

$$\underline{\mathfrak{r}}(k) = \underline{\mathfrak{r}}(k-1)\underline{\mathfrak{f}} + \underline{\underline{\mathfrak{f}}}\underline{\mathfrak{r}}(k-1) = \sum \underline{\mathfrak{f}}^i\underline{\mathfrak{r}}\underline{\underline{\mathfrak{f}}}^j,\ i+j = k-1.$$

Also, define

$$R(1) = R$$

and, for $k \ge 2$,

$$R(k) = [R(k-1),F] = [R,\underbrace{F,\dots,F}_{k-1}].$$

Set

$$\underline{\underline{a}} = \mathbb{Z}F(F'-1), \quad F' = \gamma_2(F).$$

Then we have $R(k) \le F \cap (1+\underline{\underline{r}}(k))$, $F'' \le F \cap (1+\underline{\underline{a}}^2)$ and $\gamma_\ell(F) \le F \cap (1+\underline{\underline{f}}^\ell)$. For each $1 \le k < \ell$, define

$$D(k,\ell) = F \cap (1 + \underline{\underline{r}}(k) + \underline{\underline{a}}^2 + \underline{\underline{f}}^\ell). \tag{43}$$

It follows that

$$D(k,\ell) \ge R(K)F''\ \gamma_\ell(F),$$

and the reverse inequality $D(1,\ell) \le R(1)F''\ \gamma_\ell(F)$, if it holds, would yield the most desired inequality $F \cap (1 + \mathbb{Z}F(RF''-1) + \underline{\underline{f}}^\ell) \le RF''\ \gamma_\ell(F)$. To see this, we simply observe that if $w-1 \in \mathbb{Z}(RF''-1) + \underline{\underline{f}}^\ell$ then we have, in turn, $w-1 \in \underline{\underline{r}} + \underline{\underline{a}}^2 + \underline{\underline{f}}^\ell$; $w \in D(1,\ell)$; $w \in RF''\ \gamma_\ell(F)$. We first prove,

6.1 *LEMMA*.

(a) $D(k,k+1) = R(k)F''\ \gamma_{k+1}(F)$ for all $k \ge 1$;

(b) $D(k,k+\ell)^{C(\ell)^\wedge} \le D(k+1,k+\ell)R(k)F''$ for all $k \ge 3$,

where writing $b(i) = \ell.c.m.\{1,\ldots,i\}$ and $C(j)^* = b(1)\ldots b(j)$, we define $C(\ell)^\wedge = C(1)^*\ldots C(\ell)^*$.

Proof. Let $w \in D(k,k+\ell)$, $k \ge 2$, $\ell \ge 1$. Then

$$w-1 \in \underline{\underline{r}}(k) + \underline{\underline{a}}^2 + \underline{\underline{f}}^{k+\ell}$$

and, in turn,

$$w-1 \in \underline{\underline{r}}(k) + \underline{\underline{f}}\underline{\underline{a}} + \underline{\underline{f}}^{k+\ell}. \tag{44}$$

In particular,

$$w-1 \in \underline{\underline{f}}\underline{\underline{a}} + \underline{\underline{f}}^k$$

and it follows by Theorem 3.3 that $w \in F''\ \gamma_k(F)$. Since $F''-1 \le \underline{\underline{f}}\underline{\underline{a}}$, we may assume that, in (44), $w \in \gamma_k(F)$. Let $u_k \in \underline{\underline{r}}(k)$. Then modulo $\underline{\underline{r}}(k+1)$, u_k is a $\mathbb{Z}$-linear sum of elements of the form

$$(y_1-1)\ \ldots\ (y_{t-1}-1)(r-1)(y_t-1)\ \ldots\ (y_{k-1}-1), \quad t \ge 1,$$

where $y_i \in \{x_1,\ldots,x_m\}$, $r \in \{x_1^{e(1)}\xi_1,\ldots,x_m^{e(m)}\xi_m,\xi_{m+1},\ldots,\xi_s\}$, by (41).

Set

$$\underline{\underline{x}}(k) = \mathbb{Z}F(R(k)-1).$$

We prove that if $w-1 \equiv u_k \bmod (\underline{\underline{\mathfrak{r}}}(k+1) + \underline{\underline{\mathfrak{f}\mathfrak{a}}} + \underline{\underline{\mathfrak{f}}}^{k+\ell})$ then $C(\ell)^* u_k \equiv 0 \bmod (\underline{\underline{\mathfrak{x}}}(k) + \underline{\underline{\mathfrak{r}}}(k+1) + \underline{\underline{\mathfrak{f}\mathfrak{a}}} + \underline{\underline{\mathfrak{f}}}^{k+\ell})$.

Since $w-1$ is a Lie element, it suffices to prove that $C(\ell)^* u_k$ is congruent to a $\mathbb{Z}$-linear sum of elements of the form

$$(x_{i(1)}-1) \ldots (x_{i(k)}-1)$$

with $i(1) \leq \ldots \leq i(k)$. To achieve this, we shall need to employ the following elementary congruences:

$$(y_1-1)(y_2-1) \ldots (y_k-1) \equiv (y_1-1)(y_{2\sigma}-1) \ldots (y_{k\sigma}-1) \qquad (45)$$

mod $\underline{\underline{\mathfrak{f}\mathfrak{a}}}$, where σ is any permutation of $\{2,\ldots,k\}$;

$$(r-1)(y_2-1)(y_3-1) \ldots (y_k-1)$$
$$\equiv (y_2-1)(r-1)(y_3-1) \ldots (y_k-1)$$
$$+ ([r,y_2,\ldots,y_k] - 1) \bmod (\underline{\underline{\mathfrak{r}}}(k+1) + \underline{\underline{\mathfrak{f}\mathfrak{a}}})$$

which yields

$$(r-1)(y_2-1)(y_3-1) \ldots (y_k-1)$$
$$\equiv (y_2-1)(r-1)(y_3-1) \ldots (y_k-1) \qquad (46)$$

mod $(\underline{\underline{\mathfrak{x}}}(k) + \underline{\underline{\mathfrak{r}}}(k+1) + \underline{\underline{\mathfrak{f}\mathfrak{a}}})$;

$$(y-1)(x_i^{e(i)}\xi_i-1) \equiv (y-1)(x_i^{e(i)}-1) \bmod \underline{\underline{\mathfrak{f}\mathfrak{a}}} \qquad (47)$$

$$(y-1)(\xi_j-1) \equiv 0 \bmod \underline{\underline{\mathfrak{f}\mathfrak{a}}}. \qquad (48)$$

$$(y_1-1)(y_2-1)(r-1)(y_4-1) \ldots (y_k-1)$$
$$\equiv (y_2-1)(y_1-1)(r-1)(y_4-1) \ldots (y_k-1)$$
$$+ ([y_1,y_2,r,y_4,\ldots,y_k]-1) \bmod (\underline{\underline{\mathfrak{r}}}(k+1) + \underline{\underline{\mathfrak{f}\mathfrak{a}}})$$

which yields

$$(y_1-1)(y_2-1)(r-1)(y_4-1) \ldots (y_k-1)$$
$$\equiv (y_2-1)(y_1-1)(r-1)(y_4-1) \ldots (y_k-1) \qquad (49)$$

mod $(\underline{\underline{\mathfrak{x}}}(k) + \underline{\underline{\mathfrak{r}}}(k+1) + \underline{\underline{\mathfrak{f}\mathfrak{a}}})$, $k \geq 3$.

Using (45) - (49) it follows that if $u_k \in \underline{\underline{\mathfrak{r}}}(k)$ then, modulo $\underline{\underline{\mathfrak{x}}}(k) + \underline{\underline{\mathfrak{r}}}(k+1) + \underline{\underline{\mathfrak{f}\mathfrak{a}}}$, u_k is a $\mathbb{Z}$-linear sum of elements of the form

$$z_1 = (x_{i(1)}-1) \ldots (x_{i(t)}^{e(i(t))}-1) \ldots (x_{i(k)}-1),$$

$$t \geq 2,\ i(1) \leq \ldots \leq i(k)$$

and

$$z_2 = (x_{i(j)}-1)(x_{i(1)}^{e(i(1))}-1)(x_{i(2)}-1) \ldots$$

$$\ldots \widehat{(x_{i(j)}-1)} \ldots (x_{i(k)}-1),\ j \geq 2,$$

$i(1) \leq \ldots \leq i(k)$, where $\widehat{\ }$ indicates a missing term.

For $k \geq 3$, modulo $\mathfrak{x}(k) + \mathfrak{r}(k+1) + \mathfrak{f}\mathfrak{a} + \mathfrak{f}^{k+\ell}$, we have

$$C(\ell)^* z_2 \equiv C(\ell)^* e(i(1))(x_{i(j)}-1)(x_{i(1)}-1)(x_{i(2)}-1)$$

$$\ldots \widehat{(x_{i(j)}-1)} \ldots (x_{i(k)}-1),$$

by Proposition 3.8

$$\equiv \begin{cases} C(\ell)^*(e(i(1))/e(i(2)))(x_{i(j)}-1)(x_{i(1)}-1) \\ \quad (x_{i(2)}^{e(i(2))}-1) \ldots \widehat{(x_{i(j)}-1)} \ldots (x_{i(k)}-1) \\ \qquad \text{if } j \neq 2, \text{ by Proposition 3.8} \\ C(\ell)^*(e(i(1))/e(i(3)))(x_{i(j)}-1)(x_{i(1)}-1) \\ \quad (x_{i(3)}^{e(i(3))}-1) \ldots (x_{i(k)}-1) \\ \qquad \text{if } j = 2, \text{ by Proposition 3.8,} \end{cases}$$

which by (49), (47) and (45) reduces to the form $C(\ell)^* z_1$. Thus $C(\ell)^* u_k$ is a $\mathbb{Z}$-linear sum of elements of the form $C(\ell)^* z_1$ which by Proposition 3.8 is congruent, modulo $\mathfrak{x}(k) + \mathfrak{r}(k+1) + \mathfrak{f}\mathfrak{a} + \mathfrak{f}^{k+\ell}$, to a $\mathbb{Z}$-linear sum of elements of the form $(x_{i(1)}-1) \ldots (x_{i(k)}-1)$ with $i(1) \leq \ldots \leq i(k)$, as required.

Since $\mathfrak{f}\mathfrak{x}(k) \leq \mathfrak{r}(k+1)$, it follows that

$$w^{C(\ell)^*}-1 \equiv h_k-1 \bmod (\mathfrak{r}(k+1) + \mathfrak{f}\mathfrak{a} + \mathfrak{f}^{k+\ell}) \tag{50}$$

for some $h_k \in R(k)$. Let $w_1 = w^{C(\ell)^*} h_k^{-1}$. Then $w_1-1 \in \mathfrak{r}(k+1) + \mathfrak{f}\mathfrak{a} + \mathfrak{f}^{k+1+\ell-1}$ and as before by (50) there exists $h_{k+1} \in R(k+1)$ such that $w_2 = w_1^{C(\ell-1)^*} h_{k+1}^{-1}$ with $w_2-1 \in \mathfrak{r}(k+2) + \mathfrak{f}\mathfrak{a} + \mathfrak{f}^{k+2+\ell-2}$. Repeated application of this procedure yields $w_\ell-1 \in \mathfrak{f}\mathfrak{a} + \mathfrak{f}^{k+\ell}$ where $w_\ell = w_{\ell-1}^{C(1)^*} h_{k+\ell-1}^{-1} = w^{C(\ell)^\wedge} h_k^{-C(\ell-1)^\wedge} \ldots h_{k+\ell-2}^{-C(1)^\wedge} h_{k+\ell-1}^{-1}$,

$h_i \in R(i)$. By Theorem 3.3, $w_\ell \in F'' \gamma_{k+\ell}(F)$ and it follows that $w^{C(\ell)^\wedge} \in R(k)F'' \gamma_{k+\ell}(F) \le D(k+1,k+\ell)R(k)F''$. This proves part (b) and hence also part (a) for $k \ge 3$, since $C(1)^\wedge = 1$. Finally Lemma 6.1(a) for $k = 2$ follows from (46), (47) and for $k = 1$ it is trivial. □

We next prove,

6.2 *LEMMA*. (cf. Lemma 1.5(A)). Let $w \in \gamma_{k+q}(F)$, $k,q \ge 1$, be such that $w-1 \in \underline{\underline{r}}(k) + \underline{\underline{a}}^2 + \underline{\underline{f}}^{k+q+1}$. Then $w^{b(k)}-1 \equiv f_k-1$ $\mathrm{mod}(\underline{\underline{r}}(k+1) + \underline{\underline{a}}^2 + \underline{\underline{f}}^{k+q+1})$ for some $f_k \in R(k)$, where $b(k) = \ell.c.m.\{1,\dots,k\}$.

Proof. Let $G = F/RF''$ be as in (42) and let $F^* = \langle x_1,\dots,x_m, y_1,\dots,y_{s+m}\rangle$ be free on $s+2m$ generators. With $n = k+q$, let

$$\underline{\underline{A}}^* = A_0^* \oplus \dots \oplus A_n^*$$

be the free $(n+1)$-truncated $\mathbb{Z}$-algebra (as in (2)) generated by $\{a_1,\dots,a_m,b_1,\dots,b_{s+m}\}$ and let

$$\underline{\underline{A}} = A_0 \oplus \dots \oplus A_n$$

be the corresponding subalgebra of $\underline{\underline{A}}^*$ (as in (1)) generated by $\{a_1,\dots,a_m\}$. We proceed as in the proof of Lemma 1.5(A) and conclude the proof using $\theta(\underline{\underline{a}}^2+\underline{\underline{f}}^{n+1}) = \underset{\sim}{A}^2 \cap \underline{\underline{A}}$ and Lemma 1.4. □

We are now in a position to prove our main result of this section.

6.3 *THEOREM*. (Gupta 1987). Let $G = F/RF''$ be a finite p-group. Then $D(1,n+2) = RF''\gamma_{n+2}(F)$ for all $n \le 2p-3$.

Proof. Let $n \le 2p-3$. Set $H_k = R(k)F''$, $k \ge 1$; $K_\ell = \gamma_\ell(F)$, $\ell \ge 1$ and $D_{k,\ell} = D(k,\ell)$, $1 \le k < \ell$, where $R(k) = [R,\underbrace{F,\dots,F}_{k-1}]$ and $D(k,\ell)$ is as in (43). Then $\{D_{k,\ell};\ 1 \le k < \ell\}$ is a family of normal subgroups of F satisfying:

(a) $D_{k,k+1} = H_k K_{k+1}$ (by Lemma 6.1(a));

(b) $H_k K_\ell \le D_{k,\ell}$; and

(c) $D_{k,\ell+1} \le D_{k,\ell}$, for $1 \le k < \ell$.

In addition by Lemmas 6.1(b) and 6.2 we have the property:

(d) $(K_{k+q} \cap D_{k,k+q+1})^{a(k)} \leq D_{k+1,k+q+1}H_k$,

where $a(1) = 1$, $a(2) = 2$ and, for $k \geq 3$, $a(k) = \text{g.c.d.}\{b(k), C(q+1)^{\wedge}\}$. Since $q \leq n-k+1$, it follows that $a(k)$ divides $\text{g.c.d.}\{b(k), C(n-k+2)^{\wedge}\}$.

Since $n \leq 2p-3$, it follows that $a(k)$ divides $k!$ for $k \leq p-1$ and, for $p \leq k \leq n+1$, $a(k)$ divides a number whose prime factors divide $(n-k+2)!$. In each case $(a(k),p) = 1$. Thus by Theorem 1.6 it follows that

$$D_{1,n+2}^{a(1,n+1)} \leq RF'' \gamma_{n+2}(F)$$

where $a(1,n+1) = a(1)^{\binom{n}{1}} \ldots a(n)^{\binom{n}{n}}$ is co-prime to p. On the other hand, by Corollary 3.5, $D_{1,n+2}/RF'' \gamma_{n+2}(F)$ is a p-group. Thus $D_{1,n+2} = RF'' \gamma_{n+2}(F)$ as desired. □

6.4 *CONCLUDING REMARKS*. For $p = 2$, Theorem 6.3 is best possible by virtue of Example 2.1. For $p = 3$, Theorem 6.3 also follows from Gupta & Tahara (1985). For $p \geq 3$ it remains to be seen whether or not $D_{2p}(F/RF'') = \gamma_{2p}(F/RF'')$. □

Chapter V

Generalized Magnus Embeddings

In Chapter III we studied the Fox quotient rings $\mathbb{Z}F/\underline{\underline{r}}\,\underline{\underline{f}}^n$ and the associated Fox subgroups $F \cap (1+\underline{\underline{r}}\,\underline{\underline{f}}^n)$. These are particular instances of the more general quotient rings $\mathbb{Z}F/\underline{\underline{r}}_1\cdots\underline{\underline{r}}_n$ and their associated subgroups $F \cap (1+\underline{\underline{r}}_1\cdots\underline{\underline{r}}_n)$ of the free group F where $\underline{\underline{r}}_i = \mathbb{Z}F(R_i-1)$, $R_i \trianglelefteq F$, $i = 1,\ldots,n$. It turns out that these quotient rings too admit faithful matrix representations over suitable polynomial rings and these, in turn, yield faithful representations of the groups $F/F \cap (1+\underline{\underline{r}}_1\cdots\underline{\underline{r}}_n)$. In this chapter we study these generalized embeddings and discuss some of their properties.

1. A FAITHFUL MATRIX REPRESENTATION OF $\mathbb{Z}F/\underline{\underline{r}}\,\underline{\underline{x}}$ $(\underline{\underline{x}} \leq \underline{\underline{f}})$

Let F be a free group with basis X and R be a normal subgroup of F with a free basis Y consisting of certain reduced words in F. As in I.1 we consider

$$\hat{\Omega} = \sum_{X} (\mathbb{Z}F/\underline{\underline{r}})\lambda_x(\mathbb{Z}F/\underline{\underline{x}})$$

to be the free $(\mathbb{Z}F/\underline{\underline{r}}, \mathbb{Z}F/\underline{\underline{x}})$-bimodule with a free basis $\{\lambda_x;\ x \in X\}$, where $\underline{\underline{r}} = \mathbb{Z}(R-1)$ and $\underline{\underline{x}}$ is a two-sided ideal of $\mathbb{Z}F$ contained in $\underline{\underline{f}}$. Let

$$M = \begin{bmatrix} \mathbb{Z}F/\underline{\underline{r}} & \hat{\Omega} \\ 0 & \mathbb{Z}F/\underline{\underline{x}} \end{bmatrix}$$

be the ring of 2×2 matrices and

$$\psi^*\colon \mathbb{Z}F \to M \tag{1}$$

be the ring homomorphism induced by the group homomorphism

$$\psi\colon x \to \begin{bmatrix} x+\underline{\underline{r}} & \lambda_x \\ 0 & x+\underline{\underline{x}} \end{bmatrix}, \quad x \in X, \tag{2}$$

of F into M. As before we denote by $\alpha_{ij}(u)$ the ij-entry of the matrix $\psi^*(u)$, $u \in \mathbb{Z}F$, so that

$$\psi^*(u) = \begin{bmatrix} \alpha_{11}(u) & \alpha_{12}(u) \\ 0 & \alpha_{22}(u) \end{bmatrix}. \tag{3}$$

From I.1 we record the following,

1.1 *LEMMA*. (i) $\alpha_{11}(u) = 0$ if and only if $u \in \underline{\underline{r}}$;

(ii) $\alpha_{22}(u) = 0$ if and only if $u \in \underline{\underline{x}}$;

(iii) If $w = x_1^{\varepsilon_1} \ldots x_\ell^{\varepsilon_\ell}$, $x_i \in X$, $\varepsilon_i = \pm 1$, is a reduced word of length $\ell \geq 1$ then

$$\alpha_{12}(w) = \sum_{j=1}^{\ell} \varepsilon_j (a_j R) \lambda_{x_j} (b_j + \underline{\underline{x}}/\underline{\underline{x}}),$$

where

$$a_j = x_1^{\varepsilon_1} \ldots x_{j-1}^{\varepsilon_{j-1}} x_j^{(\varepsilon_j - 1)/2} \quad \text{and} \quad b_j = x_j^{(\varepsilon_j - 1)/2} x_{j+1} \ldots x_\ell$$

are respectively the initial and terminal segments of w. □

We choose Y to be a Nielsen-reduced basis for R and denote by $|w|$ the length of a reduced word w. By Proposition I.2.4 of Lyndon & Schupp (1977), we have the following properties of Y,

1.2 *LEMMA*. Let $y \in Y$ be of the form $y = fx^\varepsilon g$, $x \in X$, $\varepsilon = \pm 1$, such that either $|f| = |g|$ or $|f| = |x^\varepsilon g|$. If $w \in F$ is a reduced word of the form $fx^\varepsilon g_1$ or $g^{-1}x^{-\varepsilon} f_1$ with $|g_1| \leq |g|$, $|f_1| \leq |f|$, then $w \in Y$ if and only if $w = y$. □

We can now prove,

1.3 *LEMMA*. Let Y be a Nielsen-reduced basis for R and $\underline{\underline{x}} \leq \underline{\underline{f}}$ be any ideal of $\mathbb{Z}F$. Then the set $\{\alpha_{12}(y);\ y \in Y\}$ is right $\mathbb{Z}F/\underline{\underline{x}}$- linearly independent.

Proof. Let $\sum_{y \in Y} \alpha_{12}(y) v_y = 0$, $v_y \in \mathbb{Z}F/\underline{\underline{x}}$, where all but finitely many of the v_y are zero. Then for $y \in Y$ of <u>maximal</u> length occurring in the above equation with $v_y \neq 0$, we write $y = fx^\varepsilon g$, $x \in X$, $\varepsilon = \pm 1$, where $|f| = |g|$ or $|f| = |x^\varepsilon g|$. By Lemma 1.1 (iii), $\alpha_{12}(y)$ has a component

$$\varepsilon (f x^{(\varepsilon-1)/2} R) \lambda_x (x^{(\varepsilon-1)/2} g + \underline{\underline{x}}/\underline{\underline{x}})$$

which lies in

$$(fx^{(\varepsilon-1)/2}R)\lambda_x(\mathbb{Z}F/\underline{\underline{\chi}}) \; (= (g^{-1}x^{(-\varepsilon-1)/2}R)\lambda_x(\mathbb{Z}F/\underline{\underline{\chi}})). \tag{4}$$

Thus for some $y_1 \in Y$ with $|y_1| \le |y|$ and $v_{y_1} \ne 0$ we must have y_1 of the form $fx^{\varepsilon}g_1$ or $g^{-1}x^{-\varepsilon}f_1$ with $|g_1| \le |g|$, $|f_1| \le |f|$, in order that $\alpha_{12}(y_1)$ contributes a component satisfying (4). This, however, is not possible by Lemma 1.2. Thus $v_y = 0$, contrary to our assumption. □

We next prove our main result of this section.

1.4 *THEOREM*. (cf. Lewin 1974). Let $\psi^*: \mathbb{Z}F \to M$ be as defined by (1). Then $\ker \psi^* = \underline{\underline{\mathfrak{r}}}\underline{\underline{\chi}}$.

Proof. As in the proof of Theorem I.1.4 the inequality $\underline{\underline{\mathfrak{r}}}\underline{\underline{\chi}} \le \ker \psi^*$ follows readily. For the reverse inequality, let $u \in \ker \psi^*$. Since $\alpha_{11}(u) = 0$ it follows that $u \in \underline{\underline{\mathfrak{r}}}$ (Lemma 1.1(i)). Thus by Proposition I.1.12, we may write

$$u = \sum_{y \in Y} (y-1)u_y, \quad u_y \in \mathbb{Z}F, \tag{5}$$

where all but finitely many of the u_y are zero. Since $\psi^*(u) = 0$, using the algebra of matrices, we obtain in turn,

$$\alpha_{12}(u) = 0;$$

$$\sum_y \alpha_{11}(y-1)\alpha_{12}(u_y) + \alpha_{12}(y-1)\alpha_{22}(u_y) = 0;$$

$$\sum_y \alpha_{12}(y)\alpha_{22}(u_y) = 0.$$

By Lemma 1.3 it follows that $\alpha_{22}(u_y) = 0$ for all y and consequently $u_y \in \underline{\underline{\chi}}$ (Lemma 1.1(ii)). By (5) it now follows that $u \in \underline{\underline{\mathfrak{r}}}\underline{\underline{\chi}}$ as was to be proved. □

1.5 *REMARKS*. By interpreting $\hat{\Omega} = \sum_x (\mathbb{Z}F/\underline{\underline{\mathfrak{r}}})\lambda_x(\mathbb{Z}F/\underline{\underline{\chi}})$ as embedded in the polynomial ring $\Omega_2 = (\mathbb{Z}F/\underline{\underline{\mathfrak{r}}} \times \mathbb{Z}F/\underline{\underline{\chi}})[\lambda_x; x \in X]$, we have shown that the kernel of the homomorphism $\mathbb{Z}F \to \Omega_2$ induced by

$$x \to \begin{bmatrix} x+\underline{\underline{\mathfrak{r}}} & \lambda_{12}^{(x)} \\ 0 & x+\underline{\underline{\chi}} \end{bmatrix}$$

is $\underline{\underline{\mathfrak{r}}}\underline{\underline{\chi}}$. □

2. A FAITHFUL MATRIX REPRESENTATION OF $\mathbb{Z}F/\mathfrak{r}_1\cdots\mathfrak{r}_n$

Let $R_1,\ldots,R_n$ be normal subgroups of a free group F with basis X and $\mathfrak{r}_i = \mathbb{Z}F(R_i-1)$, $i = 1,\ldots,n$. Further, let

$$\Omega_n = (\mathbb{Z}F/\mathfrak{r}_1 \times \ldots \times \mathbb{Z}F/\mathfrak{r}_n)[\lambda_{ij}^{(x)} \mid 1 \le i \le j \le n,\ x \in X]$$

be the polynomial ring over $(\mathbb{Z}F/\mathfrak{r}_1 \times \ldots \times \mathbb{Z}F/\mathfrak{r}_n)$ in the independent indeterminates $\lambda_{ij}^{(x)}$, $x \in X$, and

$$M_n = \begin{bmatrix} \mathbb{Z}F/\mathfrak{r}_1 & \Omega_n & & \Omega_n \\ & \mathbb{Z}F/\mathfrak{r}_2 & \cdots & \Omega_n \\ \vdots & & & \\ 0 & 0 & \cdots & \mathbb{Z}F/\mathfrak{r}_n \end{bmatrix}$$

be the ring of $n \times n$ upper-triangular matrices. Define

$$\psi_n^*\colon \mathbb{Z}F \to M_n$$

to be the linear extension of the homomorphism $\psi_n\colon F \to U(M_n)$ given by

$$\psi_n\colon x \to \begin{bmatrix} x+\mathfrak{r}_1 & \lambda_{12}^{(x)} & \lambda_{13}^{(x)} & \cdots & \lambda_{1n}^{(x)} \\ 0 & x+\mathfrak{r}_2 & \lambda_{23}^{(x)} & \cdots & \lambda_{2n}^{(x)} \\ \vdots & \vdots & \vdots & & \\ 0 & 0 & 0 & \cdots & x+\mathfrak{r}_n \end{bmatrix}.$$

Then we prove,

2.1 *THEOREM*. (Lewin 1974, Yunus 1984). Ker $\psi_n^* = \mathfrak{r}_1\cdots\mathfrak{r}_n$.

Proof. By induction on $n \ge 2$. For $n = 2$ the result follows by Remark 1.5. Let $n \ge 3$ and assume the result for $n-1$. We first prove that $\mathfrak{r}_1\cdots\mathfrak{r}_n \le \ker \psi_n^*$. Let $u = u_1\cdots u_n \in \mathbb{Z}F$ with $u_i \in \mathfrak{r}_i$. Then $\psi_n^*(u_i)$ is an $n \times n$ upper-triangular matrix whose ii-entry is zero. It follows that $\psi_n^*(u) = \psi_n^*(u_1)\cdots\psi_n^*(u_n)$ is the zero matrix and $u \in \ker \psi_n^*$.

For the reverse inclusion $\ker \psi_n^* \le \mathfrak{r}_1\cdots\mathfrak{r}_n$, let $M_{n,\hat{2}}$ be the ring of $n-1 \times n-1$ matrices obtained from M_n by deleting the 2nd rows and 2nd columns, and let

$$\psi^*_{n,\hat{2}}: \mathbb{Z}F \to M_{n,\hat{2}}$$

be the induced ring homomorphism.

Then by the induction hypothesis,

$$\ker \psi^*_{n,\hat{2}} = \underline{\underline{r}}_1 \underline{\underline{r}}_3 \cdots \underline{\underline{r}}_n .$$

Now, let $z \in \ker \psi^*_n$. Since $\ker \psi^*_n \le \ker \psi^*_{n,\hat{2}} = \underline{\underline{r}}_1 \underline{\underline{r}}_3 \cdots \underline{\underline{r}}_n$, we may assume that z is a finite sum of the form

$$z = \sum_y (y-1)u_y ,$$

where $y \in Y$, a Nielsen-reduced basis for R_1, and $u_y \in \underline{\underline{r}}_3 \cdots \underline{\underline{r}}_n \cap \underline{\underline{r}}_2$ (since $\ker \psi^*_n \le \underline{\underline{r}}_1 \underline{\underline{r}}_2$ also).

Let $\alpha_{ij}(u)$ denote the ij-entry of $\psi^*_n(u)$. Then $\alpha_{\ell,n}(u_y) = 0$ for $\ell = 3,\ldots,n$; and consequently we have, in turn,

$$0 = \alpha_{1n}(z) = \sum_y \sum_{\ell=1}^{n} \alpha_{1\ell}(y-1)\alpha_{\ell n}(u_y)$$

$$= \sum_y \alpha_{12}(y-1)\alpha_{2n}(u_y) = \sum_y \alpha_{12}(y)\alpha_{2n}(u_y) ,$$

since $\alpha_{11}(y-1) = 0$. Since Y is Nielsen-reduced, by Lemma 1.3, the set $\{\alpha_{12}(y);\ y \in Y\}$ is right $\mathbb{Z}F/\underline{\underline{x}}$-linearly independent with respect to any ideal $\underline{\underline{x}} \le \underline{\underline{f}}$. Since $\alpha_{2n}(u_y) \in \Omega_{n,\hat{1}}$, where

$$\Omega_{n,\hat{1}} = (\mathbb{Z}F/\underline{\underline{r}}_2 \times \cdots \times \mathbb{Z}F/\underline{\underline{r}}_n)[\lambda_{ij}^{(x)} \mid 2 \le i < j \le n,\ x \in X],$$

it follows by Lemma 1.3 that

$$\alpha_{2n}(u_y) = 0$$

for all $y \in Y$. Similarly, $\alpha_{2m}(u_y) = 0$ for $m = 2,\ldots,n$. It follows that $u_y \in \ker \psi^*_{n,\hat{1}} = \underline{\underline{r}}_2 \cdots \underline{\underline{r}}_n$, by the induction hypothesis. Thus $z \in \underline{\underline{r}}_1 \cdots \underline{\underline{r}}_n$ as was to be proved. □

[Theorem 2.1 answers a question of C.K. Gupta (1978).]

2.2 <u>*COROLLARY*</u>. Let $F(\underline{\underline{r}}_1 \cdots \underline{\underline{r}}_n) = F \cap (1+\underline{\underline{r}}_1 \cdots \underline{\underline{r}}_n)$. Then $F/F(\underline{\underline{r}}_1 \cdots \underline{\underline{r}}_n) \cong \langle \psi_n(x) \mid x \in X \rangle$, the group generated by $n \times n$ upper-triangular matrices $\psi_n(x)$ defined by

$$\psi_n(x) = \begin{bmatrix} xR_1 & \lambda_{12}^{(x)} & \cdots & \lambda_{1n}^{(x)} \\ 0 & xR_2 & \cdots & \lambda_{2n}^{(x)} \\ \vdots & \vdots & & \vdots \\ 0 & 0 & & xR_n \end{bmatrix}. \qquad \square$$

2.3 *REMARKS*. With $R_1 = R$, $R_i = F$, $i = 2,\ldots,n$, Theorem 2.1 and Corollary 2.2 coincide with Theorem III.1.1 and Corollary III.1.2. With $R_i \in \{R,F\}$, (i.e., $\mathfrak{r}_i \in \{\mathfrak{r},\mathfrak{f}\}$), these results were also obtained in Gupta & Gupta (1981). Since $F \cap (1+\mathfrak{r}^n) = \gamma_n(R)$, with $R_i = R$, $i = 1,\ldots,n$, Corollary 2.2 yields a faithful matrix representation of $F/\gamma_n(R)$ as in Gupta & Gupta (1978). Identification of the subgroups $F \cap (1+\mathfrak{r}_1\ldots\mathfrak{r}_n)$, $n \geq 3$, appears to be a difficult problem even when n is small and even when $\mathfrak{r}_i \in \{\mathfrak{r},\mathfrak{f}\}$. For $n = 3,4$, some progress is reported by C.K. Gupta (1978), (1983, (see Section 4). When $R_i \in \{F,F'\}$, Corollary 2.2 yields a family of relatively free linear groups $F/F(\mathfrak{r}_1\ldots\mathfrak{r}_n)$, contributing towards a problem of Wehrfritz (1973), page 34. $\square$

3. RESIDUAL NILPOTENCE OF $F/F \cap (1+\mathfrak{r}_1\ldots\mathfrak{r}_n)$

In this section we prove some important generalizations of the Passi-Lichtman Theorem II.4.2, developed by Yunus (1984) and C.K. Gupta & Passi (1984). Among other things, we prove that if $\Delta^\omega(F/R) = 0$, then $\Delta^\omega(F/F \cap (1+\mathfrak{r}_1\ldots\mathfrak{r}_n)) = 0$ where $\mathfrak{r}_i \in \{\mathfrak{r},\mathfrak{f}\}$, $\mathfrak{r} = \mathbb{Z}F(R-1)$. The exposition here is due to C.K. Gupta & Passi.

Let G be a group and $KG[\Lambda]$ be the polynomial ring in the set Λ of commuting indeterminates over the group ring KG, where K is a commutative ring with 1_K (its identity). For each $n \geq 2$, let

$$M_{n,K}(G) = \{\underset{\sim}{u} = [u_{ij}]_{n\times n} \mid u_{ij} \in KG[\Lambda],\ u_{ij} = 0 \text{ for } i > j,\ u_{ii} \in G\}$$

be the group of $n \times n$ upper-triangular matrices with diagonal entries in G. Let $M_{n,K}(KG[\Lambda])$ be the ring of all upper-triangular matrices over $KG[\Lambda]$. Then the inclusion map

$$\eta\colon M_{n,K}(G) \to M_{n,K}(KG[\Lambda]),\quad \underset{\sim}{u} \to \underset{\sim}{u}$$

entends by linearity to a ring homomorphism

$$\eta^*: KM_{n,K}(G) \to M_{n,K}(KG[\Lambda])$$

and the matrix multiplication yields, for $m \geq 1$,

$$\eta^*(\Delta_K^{n+m}(M_{n,K}(G)) \leq M_{n,K}(\Delta_K^m(G)[\Lambda],$$

where $\Delta_K(H)$ is the augmentation ideal of KH. The ring homomorphism $\mathbb{Z} \to K$, $m \to m1_K$, induces a ring homomorphism

$$\alpha: KM_{n,\mathbb{Z}}(G) \to KM_{n,K}(G).$$

Setting $\eta^{**} = \eta^*\alpha$ yields

3.1 *LEMMA*. (cf. Gupta & Passi (1976). Let K be a commutative ring with 1. Then there is a homomorphism

$$\eta^{**}: M_{n,\mathbb{Z}}(G) \to M_{n,K}(KG[\Lambda]),$$

such that

$$\eta^{**}(\Delta_K^\omega(M_{n,\mathbb{Z}}(G)) \leq M_{n,K}(\Delta_K^\omega(G)[\Lambda]). \qquad \square$$

3.2 *LEMMA*. (C.K. Gupta & Passi 1983).

(a) If G is residually torsion-free nilpotent, then so is $M_{n,\mathbb{Z}}(G)$;

(b) If G is residually nilpotent p-group of bounded exponent, then so is $M_{n,\mathbb{Z}}(G)$.

Proof. For the proof of (a), let G be residually torsion-free nilpotent. Then $\Delta_Q^\omega(G) = 0$ (Passi (1979), p.90). Lemma 3.1 with $K = Q$ yields

$$\eta^{**}(\Delta_Q^\omega(M_{n,\mathbb{Z}}(G))) = 0.$$

Now if $\underset{\sim}{u} \in \bigcap_m \sqrt{\gamma_m(M_{n,\mathbb{Z}}(G))}$, then $\underset{\sim}{u}-I \in \Delta_Q^\omega(M_{n,\mathbb{Z}}(G))$ and consequently $\eta^{**}(\underset{\sim}{u}-I) = 0$ which yields $\underset{\sim}{u} = I$ and it follows that $M_{n,\mathbb{Z}}(G)$ is residually torsion-free nilpotent.

For the proof of (b), let G be residually nilpotent p-group of bounded exponent. Then $\Delta_{\mathbb{Z}'}^\omega(G) = 0$, where $\mathbb{Z}' = \mathbb{Z}_{p^s}$, $s \geq 1$ (Passi (1979), p.84). Thus by Lemma 3.1 with $K = \mathbb{Z}'$, it follows that

$$\eta^{**}(\Delta_{\mathbb{Z}'}^\omega(M_{n,\mathbb{Z}}(G))) = 0.$$

Now, if $\underset{\sim}{u} \in \bigcap_{i,j} M_{n,\mathbb{Z}}(G)^{p^i}\gamma_j(M_{n,\mathbb{Z}}(G))$, then $\underset{\sim}{u}-I \in \Delta^{\omega}_{\mathbb{Z}'}(M_{n,\mathbb{Z}}(G))$ (Passi (1979), p.96) and consequently $\eta^{**}(\underset{\sim}{u}-I) = 0$. As before, $\underset{\sim}{u} = I$ and it follows that $M_{n,\mathbb{Z}}(G)$ is residually nilpotent p-group of bounded exponent. □

3.3 *THEOREM*. (C.K. Gupta & Passi 1984). If $\Delta^{\omega}(G) = 0$ then $\Delta^{\omega}(M_{n,\mathbb{Z}}(G)) = 0$.

Proof. Let $\Delta^{\omega}(G) = 0$ then by Theorem II.4.2, there are two cases.

CASE I. (G is residually torsion-free nilpotent.) In this case, by Lemma 3.2(a), $M_{n,\mathbb{Z}}(G)$ is residually torsion-free nilpotent and, by Therem II.4.2, $\Delta^{\omega}(M_{n,\mathbb{Z}}(G)) = 0$.

CASE II. (G is discriminated by the class $\mathcal{C}$ of nilpotent groups of bounded prime power exponents.) Let $\{\underset{\sim}{u}_1,\ldots,\underset{\sim}{u}_m\}$ be a set of distinct elements of $M_{n,\mathbb{Z}}(G)$. Let S_i, $1 \le i \le m$, be the union of supports in G of elements of $\mathbb{Z}G$ which occur as coefficients in the entries of the matrices $\underset{\sim}{u}_i$ and set $S = \bigcup_{i=1}^{m} S_i$. Then S is a finite set of distinct elements $g_1,\ldots,g_t$, say. By hypothesis, there exists a group $H \in \mathcal{C}$ and a homomorphism $\alpha\colon G \to H$ such that $\alpha(g_1),\ldots,\alpha(g_t)$ are distinct elements of H. Since $\{\alpha(g_1),\ldots,\alpha(g_t)\}$ is the support of the matrices $\alpha(\underset{\sim}{u}_1),\ldots,\alpha(\underset{\sim}{u}_m)$, it follows that these matrices are distinct elements of $M_{n,\mathbb{Z}}(H)$. Thus by Lemma 3.2(b) there exists a nilpotent p-group L of bounded exponent and a homomorphism $\beta\colon M_{n,\mathbb{Z}}(H) \to L$ such that $\beta(\alpha(\underset{\sim}{u}_1)),\ldots,\beta(\alpha(\underset{\sim}{u}_m))$ are distinct elements of L. It follows that $M_{n,\mathbb{Z}}(G)$ is discriminated by the class $\mathcal{C}$ and by Theorem II.4.2, $\Delta^{\omega}(M_{n,\mathbb{Z}}(G)) = 0$. □

3.4 *THEOREM*. (C.K. Gupta & Passi 1984). If $\Delta^{\omega}(F/R) = 0$ and if $R_i \in \{R,F\}$ then $\Delta^{\omega}(F/F \cap (1+\underline{\underline{r}}_1\cdots\underline{\underline{r}}_n)) = 0$, where $\underline{\underline{r}}_i = \mathbb{Z}F(R_i-1)$.

Proof. If $\Delta^{\omega}(F/R) = 0$ and if $R_i \in \{R,F\}$, then $\Delta^{\omega}(F/R_1 \times \ldots \times F/R_n) = 0$. By Corollary 2.2, $F/F \cap (1+\underline{\underline{r}}_1\cdots\underline{\underline{r}}_n)$ is embedded in $M_{n,\mathbb{Z}}(G)$, where $G = F/R_1 \times \ldots \times F/R_n$. Since by Theorem 3.3, $\Delta^{\omega}(M_{n,\mathbb{Z}}(G)) = 0$, it follows that $\Delta^{\omega}(F/F \cap (1+\underline{\underline{r}}_1\cdots\underline{\underline{r}}_n)) = 0$. □

3.5 *THEOREM*. (Yunus 1984). If F/R_i is residually torsion-free nilpotent for $i = 1,\dots,n$, then $\Delta^{\omega}(F/F \cap (1+\mathfrak{r}_1\cdots\mathfrak{r}_n)) = 0$.

Proof. If F/R_i is residually torsion-free nilpotent for $i = 1,\dots,n$, then $F/R_1 \times \dots \times F/R_n$ is residually torstion-free nilpotent and consequently $\Delta^{\omega}(G) = 0$ with $G = F/R_1 \times \dots \times F/R_n$. Thus by Theorem 3.3, $\Delta^{\omega}(M_{n,\mathbb{Z}}(G)) = 0$ which yields $\Delta^{\omega}(F/F \cap (1+\mathfrak{r}_1\cdots\mathfrak{r}_n)) = 0$. □

3.6 *THEOREM*. (Gupta & Gupta 1981, Yunus 1984). $\mathbb{Z}F/\mathfrak{r}_1\cdots\mathfrak{r}_n$ is residually nilpotent if and only if $\mathbb{Z}F/\mathfrak{r}_i$ is residually nilpotent for each $i = 1,\dots,n$.

Proof. Let $\mathbb{Z}F/\mathfrak{r}_1\cdots\mathfrak{r}_n$ be residually nilpotent and let $u \in \bigcap_m(\mathfrak{r}_i+\mathfrak{f}^m)$ for some fixed $i = 1,\dots,n$. For each $k \neq i$, we choose a generator y_k of R_k and set

$$v = (y_1-1)\ \dots\ (y_{i-1}-1)u(y_{i+1}-1)\ \dots\ (y_n-1).$$

Then $v \in \bigcap_m(\mathfrak{r}_1\cdots\mathfrak{r}_n+\mathfrak{f}^m) = \mathfrak{r}_1\cdots\mathfrak{r}_n$, by hypothesis. Now $v \equiv 0 \bmod \mathfrak{r}_1\cdots\mathfrak{r}_n$ implies, using Proposition I.1.12, that $u \in \mathfrak{r}_i$. It follows that $\mathbb{Z}F/\mathfrak{r}_i$ is residually nilpotent for each i.

Conversely, let $\mathbb{Z}F/\mathfrak{r}_i$ be residually nilpotent for each i so that $\bigcap_m(\mathfrak{r}_i+\mathfrak{f}^m) = \mathfrak{r}_i$ for $i = 1,\dots,n$. Let $u \in \bigcap_m(\mathfrak{r}_1\cdots\mathfrak{r}_n+\mathfrak{f}^m)$ and consider the matrix $\psi_n^*(u)$. Then by Theorem 2.1 it suffices to prove that $\psi_n^*(u) = 0$ or equivalently, $\alpha_{ij}(u) = 0$ for all $1 \le i \le j \le n$. For each $1 \le i \le j \le n$ and each $m \ge 1$, we set

$$\theta_{i,j}(\mathfrak{f}^m) = \sum_{\underset{\sim}{t}} (\mathfrak{f}^{t_i}+\mathfrak{r}_i)\ \dots\ (\mathfrak{f}^{t_j}+\mathfrak{r}_j)$$

where $\underset{\sim}{t}$ ranges over all $(j-i+1)$-tuples $(t_i,\dots,t_j)$ of non-negative integers satisfying $t_i + \dots + t_j = m$. Also for $1 \le i \le n$, set $\Lambda_{i,i} = \{1\}$ and for $1 \le i < j \le n$, set

$$\Lambda_{i,j} = \{\lambda_{r,s}^{(x)};\ i \le r < s \le j,\ x \in X\}.$$

Then if $v \in \mathfrak{f}^{j-i+m}$, a simple induction on $j-i \ge 0$ shows that

$$\alpha_{ij}(v) \in \theta_{i,j}(\mathfrak{f}^m)\Lambda_{i,j}.$$

Thus, since $u \in \bigcap_m(\mathfrak{r}_1\cdots\mathfrak{r}_n+\mathfrak{f}^m) \le \bigcap_m(\mathfrak{r}_i\cdots\mathfrak{r}_j+\mathfrak{f}^{j-i+m})$, it follows that

$$\alpha_{ij}(u) \in (\bigcap_m \theta_{i,j}(\mathfrak{f}^m))\Lambda_{i,j}$$

$$\leq (\sum_{k=i}^{j} (\mathfrak{r}_i+\mathfrak{f}) \ldots (\bigcap_m (\mathfrak{r}_k+\mathfrak{f}^m)) \ldots (\mathfrak{r}_j+\mathfrak{f}))\Lambda_{i,j}$$

$$= 0, \text{ by hypothesis.}$$

This completes the proof. □

3.7 *REMARKS*. The converse of Theorem 3.4 also holds in certain cases. For instance, if $R_i \in \{F,R\}$ for $i = 1,\ldots,n$ (i.e., $\mathfrak{r}_i \in \{\mathfrak{f},\mathfrak{r}\}$) then C.K. Gupta & Passi have shown that if F/R is periodic and $(\mathfrak{r}_1,\mathfrak{r}_n) \neq (\mathfrak{f},\mathfrak{f})$ then $\Delta^\omega(F/F \cap (1+\mathfrak{r}_1\ldots\mathfrak{r}_n)) = 0$ implies $\Delta^\omega(F/R) = 0$. Similar results hold if F/R is torsion-free. In particular, if either F/R is periodic or torsion-free then $\Delta^\omega(F/R) = 0$ if and only if $\Delta^\omega(F/\gamma_n(R)) = 0$ (Yunus 1984). Kuz'min (1984) has shown that if F/R is residually torsion-free nilpotent then so is $F/F \cap (1+\mathfrak{f}\mathfrak{r}\mathfrak{f})$. In another direction, Hartley (1984) has shown that if $V(R)$ is fully invariant in R with $R/V(R)$ torsion-free nilpotent then $\Delta^\omega(F/R) = 0$ if and only if $\Delta^\omega(F/V(R)) = 0$. □

4. IDENTIFICATION OF $F \cap (1+\mathfrak{r}_1\ldots\mathfrak{r}_n)$ FOR SMALL n

By Theorem I.1.6, we have

$$F \cap (1+\mathfrak{r}_1\mathfrak{r}_2) = [R_1 \cap R_2,\ R_1 \cap R_2].$$

The identification of $F \cap (1+\mathfrak{r}_1\mathfrak{r}_2\mathfrak{r}_3)$ is much more difficult. Some of what is known is as follows:

(a) $F \cap (1+\mathfrak{f}^3) = \gamma_3(F)$ (See Theorem I.3.7)

(b) $F \cap (1+\mathfrak{r}^3) = R \cap (1 + \mathbb{Z}R(R-1)^3) = \gamma_3(R)$

(c) $F \cap (1+\mathfrak{f}^2\mathfrak{r}) = F \cap (1+\mathfrak{r}\mathfrak{f}^2) = F(2,R) = [R \cap F',\ R \cap F']$ (see Theorem III.2.1)

(d) $F \cap (1+\mathfrak{r}^2\mathfrak{f}) = F \cap (1+\mathfrak{f}\mathfrak{r}^2) = \gamma_3(R)$ (see Theorem 4.7 below)

(e) $F \cap (1+\mathfrak{r}\mathfrak{f}\mathfrak{r}) = \gamma_3(R)$ (see Corollary 4.5 below).

The case $F \cap (1+\mathfrak{f}\mathfrak{r}\mathfrak{f})$ is most interesting and has drawn considerable interest in the literature. Since $\mathfrak{f}\mathfrak{r}\mathfrak{f} \leq \mathfrak{r}\mathfrak{f}$ and $F \cap (1+\mathfrak{r}\mathfrak{f}) = R'$ it follows that $F \cap (1+\mathfrak{f}\mathfrak{r}\mathfrak{f}) \leq R'$. On the other hand, $[R',F]-1 \leq \mathfrak{r}\mathfrak{r}\mathfrak{f} + \mathfrak{f}\mathfrak{r}\mathfrak{r} \leq \mathfrak{f}\mathfrak{r}\mathfrak{f}$. It follows that $[R',F] \leq F \cap (1+\mathfrak{f}\mathfrak{r}\mathfrak{f}) \leq R'$, and it is of interest to know when $F \cap (1+\mathfrak{f}\mathfrak{r}\mathfrak{f}) = [R',F]$, in which case we have, by Corollary 2.2, a faithful matrix representation of $F/[R',F]$. The first important contribution towards the identification problem is due to C.K. Gupta (1969), who identified the subgroup $F \cap (1+\mathfrak{f}\mathfrak{a}\mathfrak{f})$, where $\mathfrak{a} = \mathbb{Z}F(F'-1)$. She proved,

4.2 *THEOREM*. (C.K. Gupta 1969, 1973). $F \cap (1+\mathfrak{f}\mathfrak{a}\mathfrak{f}) = K_4(F)[F'',F]$ where $K_4(F) \leq F''$ is the fully invariant closure of a 4-variable word w given by

$$w = [x_1, x_2^{-1}; x_3, x_4][x_1^{-1}, x_3^{-1}; x_4, x_2][x_1^{-1}, x_4^{-1}; x_2, x_3]$$

$$[x_3^{-1}, x_4^{-1}; x_1, x_2][x_4^{-1}, x_2^{-1}; x_1, x_3][x_2^{-1}, x_3^{-1}, x_1, x_4].$$

In addition, if rank $F = n$ then $K_4(F)[F'',F]/[F'',F]$ is an elementary abelian 2-group of rank $\binom{n}{4}$. In particular, if $n = 2,3$, then the free center-by-metabelian group $F/[F'',F]$ has a faithful 3×3 matrix representation given by Corollary 2.2.□

[Another significance of this theorem is that it provides the first known example of a variety defined by an outer commutator (namely, $[[x_1,x_2],[x_3,x_4],x_5]$) whose free groups are not torsion-free.]

Using homological methods Kuz'min (1977), (1982), (1984) has studied the detailed structure of $F \cap (1+\mathfrak{f}\mathfrak{r}\mathfrak{f})$ and has proved, in particular,

4.3 *THEOREM*. (Kuz'min 1984). If F/R has no element of order 2 then $F \cap (1+\mathfrak{f}\mathfrak{r}\mathfrak{f}) = [R',F]$ if and only if the homology group $H_4(G,\mathbb{Z}) = 0$. □

[The sufficiency statement of the above theorem was also proved by Thomson (1977), unpublished notes.]

A more satisfactory identification of $F \cap (1+\mathfrak{f}\mathfrak{r}\mathfrak{f})$ is given by Stöhr (1984), who proved that (see Theorem 4.8)

$$F \cap (1+\mathfrak{f}\mathfrak{r}\mathfrak{f}) = \sqrt{[R',F]}.$$

When $\mathfrak{r}_1, \mathfrak{r}_2, \mathfrak{r}_3$ are arbitrary the best known case is the following result due to C.K. Gupta (1978),

4.4 *THEOREM*. (C.K. Gupta). $F \cap (1+\mathfrak{r}\mathfrak{f}\mathfrak{s}) = [R' \cap S, R' \cap S][R \cap S', R \cap S'][R' \cap S', R \cap S]$.

Proof. We need only prove that if $w-1 \in \mathfrak{r}\mathfrak{f}\mathfrak{s}$, then $w \in H = [R' \cap S, R' \cap S][R \cap S', R \cap S'][R' \cap S', R \cap S]$. Since $\mathfrak{r}\mathfrak{f}\mathfrak{s} \le \mathfrak{r}\mathfrak{s}$, it follows by Theorem I.1.6 that $w \in [R \cap S, R \cap S]$. We need some information about the quotient $R \cap S/R' \cap S'$. Since $R' \cap S = \ker\{\theta: R \cap S \to R/R', w \to wR'\}$ it follows that $R \cap S/R' \cap S$ is free abelian. Similarly $R' \cap S/R' \cap S'$ is free abelian. It follows that $R \cap S/R' \cap S'$ is free abelian and each of $R' \cap S/R' \cap S'$ and $R \cap S'/R' \cap S'$ is a direct factor. Thus we can choose a basis $a_1, a_2, \ldots, a_1', a_2', \ldots$ for $R \cap S$ such that $a_1', a_2', \ldots$ is a basis for $R' \cap S$ (modulo $R' \cap S'$). Since $\gamma_3(R \cap S) \le H$ and $[R' \cap S, R' \cap S] \le H$, we can write w ($\in [R \cap S, R \cap S]$) as

$$w \equiv \prod_i [a_i, w_i] \pmod H$$

where $w_i \in \langle a_{i+1}, \ldots, a_1', a_2', \ldots \rangle$.

Now, $w-1 \in \mathfrak{r}\mathfrak{f}\mathfrak{s}$ implies

$$\sum_i (a_i-1)(w_i-1) - (w_i-1)(a_i-1) \equiv 0 (\mathfrak{r}\mathfrak{f}\mathfrak{s}).$$

Since w_i-1 does not involve $a_1, \ldots, a_i$, and since $a_1, a_2, \ldots$ are linearly independent modulo R', we must have, in turn,

$$(a_i-1)(w_i-1) \in \mathfrak{r}\mathfrak{f}\mathfrak{s}, \quad i = 1, \ldots, t.$$

Consequently, by I.1.12, $w_i-1 \in \mathfrak{f}\mathfrak{s}$ which yields $w_i \in S' \cap R$ for each i. This proves $w \in [R \cap S, R \cap S']$.

We next choose a basis $b_1, b_2, \ldots, b_1', b_2', \ldots$ for $R \cap S$ such that $b_1', b_2', \ldots$ are a basis for $R \cap S'$ (modulo $R' \cap S'$). Since $[R \cap S', R \cap S'] \le H$, as before we can write w as

$$w \equiv \prod_i [v_i, b_i] \pmod H$$

where $v_i = v_i(b_{i+1}, \ldots, b_1', b_2', \ldots) \in R \cap S'$.

As before,

$$w-1 \equiv \sum_i (v_i-1)(b_i-1) - (b_i-1)(v_i-1) \mod \mathfrak{r}\mathfrak{f}\mathfrak{s}$$

gives, in turn,

$$(v_i-1)(b_i-1) \in \mathfrak{r}\mathfrak{f}\mathfrak{s}, \quad i = 1, \ldots, t.$$

Consequently, $v_i-1 \in \mathfrak{r}\mathfrak{f}$ which yields $v_i \in R' \cap (R \cap S') = R' \cap S'$. Thus, modulo H, $w \in [R' \cap S', R \cap S] \le H$. This proves $w \in H$. □

4.5 *COROLLARY*. (C.K. Gupta). $F \cap (1+\mathfrak{r}\mathfrak{f}\mathfrak{r}) = \gamma_3(R)$. □

We now state some additional results which are relevant for our present discussion. The reader is referred to the appropriate source for a proof.

4.6 *THEOREM*. (C.K. Gupta 1983).

(a) $F \cap (1+\mathfrak{f}^2\mathfrak{r}^2) = \gamma_3(R \cap F')\gamma_4(R)$.

(b) $F \cap (1+\mathfrak{r}\mathfrak{f}^2\mathfrak{r}) = [R \cap F', R \cap F', R]\gamma_4(R)$. □

4.7 *THEOREM*. (Gruenberg (1970), page 53). $F \cap (1+\mathfrak{f}\mathfrak{r}^n) = F \cap (1+\mathfrak{r}^n\mathfrak{f}) = \gamma_{n+1}(R)$ for all $n \geq 1$. □

4.8 *THEOREM*. (Stöhr 1984). $F \cap (1+\mathfrak{f}\mathfrak{r}^n\mathfrak{f}) = \sqrt{[\gamma_{n+1}(R),F]}$ for all $n \geq 1$. □

[This settles a conjecture of Gupta & Gupta 1978).]

4.9 *THEOREM*. (Stöhr 1984). If F/R is finite of order coprime to $n+1$ then $F \cap (1+\mathfrak{f}\mathfrak{r}^n\mathfrak{f}) = [\gamma_{n+1}(R),F]$. (See also, Thomson 1979). □

We close this section with the following remarkable result due to Stöhr extending C.K. Gupta's result to arbitrary primes.

4.10 *THEOREM*. (Stöhr 1985). For any prime p, the torsion subgroup of $F/[\gamma_p(F'),F]$ is an elementary abelian p-group of rank $\binom{n}{4}$ where $n = \text{rank } F$. □

5. SOME APPLICATIONS OF THE GENERALIZED EMBEDDINGS

In this section we state without proofs some results which are either obtained using these matrix representations or influenced by these considerations. The reader is referred to the appropriate source for the details.

Since $F \cap (1+\mathfrak{r}^n) = \gamma_n(R)$, $F/\gamma_n(R)$ admits a faithful matrix representation as an $n \times n$ matrix group over $\Omega_n = \mathbb{Z}(F/R \times \ldots \times F/R)[\Lambda]$ and, as in II.2.1, has the solvable word problem if F/R has the solvable word problem. In particular,

5.1 *THEOREM*. Free polynilpotent groups have the solvable word problem. □

Extending II.2.4, it is proved in Gupta & Gupta (1978) that if F/R is a finitely generated recursively presented torsion-free group with the solvable power problem and the solvable conjugacy problem, and if $F/\gamma_{n-1}(R)$, $n \geq 3$, has the solvable conjugacy problem then the group $F/\gamma_n(R)$ has the solvable conjugacy problem. In particular,

5.2 *THEOREM*. (Sarkisyan 1972, Gupta & Gupta 1978). Free polynilpotent groups have the solvable conjugacy problem. □

5.3 *COROLLARY*. (Kargapolov & Remeslennikov 1966). Free solvable groups have the solvable conjugacy problem. □

Using the matrix representations of $F/[\gamma_n(R),F]$ (via $[\gamma_n(R),F] \leq F \cap (1+\mathfrak{f}\mathfrak{r}^{n-1}\mathfrak{f})$), the following results are proved.

5.4 *THEOREM*. (Gupta & Gupta 1978). The centre of $F/[\gamma_n(R),F]$ is precisely $\gamma_n(R)/[\gamma_n(R),F]$, $n \geq 2$. □

5.5 *THEOREM*. (Gupta & Gupta 1978, Baumslag, Strebel & Thomson 1980).

(a) If F/R is infinite then $F/\gamma_n(R)$, $n \geq 2$, cannot be finitely presented;

(b) Free non-nilpotent polynilpotent groups cannot be finitely presented. □

Using Theorem II.3.5 a result of Mital & Passi (1973) yields,

5.6 *THEOREM*. (Mital & Passi). If $R \trianglelefteq F$ and rank $F \geq 2$, then $\gamma_n(R)/\gamma_{n+1}(R)$ is a faithful $\mathbb{Z}(F/R)$-module for all $n \geq 1$. □

Now we state certain results from Gupta, Laffey & Thomson (1979) which are influenced by the matrix representations.

5.7 *THEOREM*. (Gupta, Laffey & Thomson). Let G be a finite group given by a free presentation

$$1 \to R \to F \to G \to 1$$

with rank $F = j \geq 2$. Then

(a) The character χ_k of the k-th relation module $(\gamma_k(R)/\gamma_{k+1}(R)) \otimes_{\mathbb{Z}} \mathbb{Q}$, $k \geq 1$, is given by the formula

$$\chi_k(g) = \frac{1}{k} \sum_{d|k} \mu(d)(\chi(g^d))^{k/d}, \quad g \in G,$$

where

$$\chi(g) = \begin{cases} 1 + |G|(j-1) & \text{if } g = 1 \\ 1 & \text{if } g \neq 1 \end{cases}$$

is the character χ of $R/R' \otimes_{\mathbb{Z}} \mathbb{Q}$ and μ is the Möbius function;

(b) Let $C_k/\gamma_{k+1}(R)$ be the centre of $F/\gamma_{k+1}(R)$, $k \geq 1$. Then the rank r_k of $C_k/\gamma_{k+1}(R)$ is given by the formula

$$r_1 = j \quad (= \text{rank } F),$$

$$r_k = \frac{1}{k|G|} \sum_{d|k} \mu(d)\nu_d(G)(m^{k/d}-1), \; k \geq 2,$$

where $m = 1 + (j-1)|G| = \text{rank } R$ and $\nu_d(G)$ is the number of elements g of G with $g^d = 1$;

(c) If $(|G|,k) = 1$, $k \geq 2$, then the rank r_k of $C_k/\gamma_{k+1}(R)$ is $m(k)/|G|$, where $m(k)$ is the rank of $\gamma_k(R)/\gamma_{k+1}(R)$. □

Finally, we state the following extension of Theorem II.1.13,

5.8 *THEOREM*. (Gupta, Laffey & Thomson 1979). If F/R is a non-trivial finite group then, for each $k \geq 1$, the second centre of $F/\gamma_{k+1}(R)$ coincides with its centre. □

5.9 *REMARKS*. In view of II.1.10, it would be desirable to have an explicit description of $C_k/\gamma_{k+1}(R)$ for $k \geq 2$. See Stöhr (1986). □

6. THE PROBLEM SECTION

6.1 *PROBLEM*. In view of the successful solution of the Fox problem for free groups, it is natural to raise the same problem for arbitrary groups. In the language of free group rings this amounts to the problem of identification of $F \cap (1+\mathfrak{s}+\mathfrak{r}\mathfrak{f}^n)$, $n \geq 1$, where $\mathfrak{s} = \mathbb{Z}F(S-1)$, $\mathfrak{r} = \mathbb{Z}F(R-1)$, $R,S \trianglelefteq F$ with $S \leq R$. When $n = 1$ the solution is given by Corollary I.1.15 where it is proved that $F \cap (1+\mathfrak{s}+\mathfrak{r}\mathfrak{f}) = SR'$. When $n = 2$ some progress is reported in Khambadkone (1985). □

6.2 *PROBLEM*. Another problem suggested by the identification of $F \cap (1+\mathfrak{r}\mathfrak{f}^n)$ is to seek identification of some other instances of $F \cap (1+\mathfrak{r}_1\cdots\mathfrak{r}_n)$, $n \geq 3$, where $\mathfrak{r}_i = \mathbb{Z}F(R_i-1)$, $R_i \trianglelefteq F$. Some particular cases of interest are: $F \cap (1+\mathfrak{f}^i\mathfrak{r}^j\mathfrak{f}^k)$, $i+j+k \geq 4$. The same question with $\mathfrak{r} = \mathfrak{a}_c = \mathbb{Z}F(\gamma_c(F)-1)$, $c \geq 2$, is equally attractive. □

6.3 *PROBLEM*. Give a complete description of $F \cap (1+\mathfrak{r}\mathfrak{s}\mathfrak{t})$. □

6.4 *PROBLEM*. The identification of $F \cap (1+\mathfrak{f}\mathfrak{a}+\mathfrak{f}^n) = F''\gamma_n(F)$ (Theorem IV.3.3) and, more generally, of $F \cap (1+\mathfrak{f}\mathfrak{a}_c+\mathfrak{f}^n) = [\gamma_c(F),\gamma_c(F)]\gamma_n(F)$, $c \geq 2$, due to Gupta, Gupta & Levin (1986) suggest the problem of identifying $F \cap (1+\mathfrak{f}\mathfrak{s}+\mathfrak{f}^n)$, $\mathfrak{s} = \mathbb{Z}F(S-1)$ with F/S finite nilpotent. The same problem with F/S finite abelian and n small still awaits solution (see Theorem IV.4.1). □

6.5 *PROBLEM*. Another problem suggested by the identification of $F \cap (1+\mathfrak{f}\mathfrak{a}_c+\mathfrak{f}^n)$ is to seek identification of $F \cap (1+\mathfrak{f}\mathfrak{a}_c\mathfrak{f})$, $c \geq 2$. When $c = 2$, a solution has been given by C.K. Gupta & Levin (1986). □

6.6 *PROBLEM*. Example IV.2.1 shows that in general $F \cap (1+\mathfrak{r}+\mathfrak{f}^4) \neq R\gamma_4(F)$. For $n \geq 5$ no example is known with $F \cap (1+\mathfrak{r}+\mathfrak{f}^n) \neq R\gamma_n(F)$. □

6.7 *PROBLEM*. The difficulty in the solution of the dimension subgroup problem suggests the following weaker question: Given $n \geq 4$, does there exist m $(= m(n))$ such that $D_m(G) \leq \gamma_n(G)$ for all groups G? The same question remains open when G is restricted to the class of metabelian groups. □

6.8 *PROBLEM*. If G is a finitely generated metabelian group then Theorem IV.4.3 shows that for some sufficiently large n_0 $(= n_0(G/G'))$, $D_n(G) = \gamma_n(G)$ for all $n \geq n_0$. We are tempted to ask the same question for arbitrary finitely generated groups G: Does there exist n_0 $(= n_0(G))$ such that $D_n(G) = \gamma_n(G)$ for all $n \geq n_0$? □

6.9 *PROBLEM*. Other weaker versions of the dimension subgroup problem are: (a) Is $F \cap (1+\mathfrak{r}^k+\mathfrak{f}^n) \leq R\gamma_n(F)$ for some $k \leq n-2$? (b) Is $F \cap (1+\mathfrak{r}(k)+\mathfrak{f}^n) \leq R\gamma_n(F)$ for some $k \leq n-2$? [Here $\mathfrak{r}(k) = \sum_{i+j=k-1} \mathfrak{f}^i\mathfrak{r}\mathfrak{f}^j$.] □

6.10 *PROBLEM*. If G is metabelian then Corollary IV.3.7 yields an interesting fact that $D_n(G)/\gamma_n(G)$ is abelian for all n. Is this a general phenomenon? Otherwise, we ask the following weaker question: Given G, does there exist ℓ $(= \ell(G))$ such that, for all n, $D_n(G)/\gamma_n(G)$ is solvable of length at most ℓ? [Note that an affirmative answer to Problem 6.9(a) would yield an affirmative answer to the latter question.] □

Bibliography

1935

Magnus, Wilhelm, "Beziehungen swischen Gruppen und Idealen in einem speziellen Ring", Math. Ann. 111 (1935), 259-280.

Schumann, H.G., "Über Modulun und Gruppenbilder", Math. Ann. 114 (1935), 385-413.

1936

Grün, Otto, "Über die Faktorgruppen freier Gruppen I", Deutsche Math. (Jahrgang 1), 6 (1936), 772-782.

1937

Magnus, Wilhelm, "Über Beziehungen zwischen höheren Kommutatoren", J. reine angew. Math. 177 (1937), 105-115.

Witt, E., "Treue Darstellung Liescher Ringe", J. reine angew. Math. 177 (1937), 152-160.

1939

Magnus, Wilhelm, "On a theorem of Marshall Hall", Ann. of Math. (Ser. II) 40 (1939), 764-768.

1953

Fox, Ralph, H., "Free differential calculus I - Derivations in free group rings", Ann. of Math. 57 (1953), 547-560.

1954

Gaschutz, Wolfgang, "Über modulare Darstellungen endlicher Gruppen, die von freien Gruppen induziert werden", Math. Zeitchr. 60 (1954), 274-286.

Lazard, Michel, "Sur les groupes nilpotents et les asseaux de Lie", Ann. Sci. Ecole Norm. Sup. (3) 71 (1954), 101-190.

1955

Auslander, Maurice and Lyndon, R.C., "Commutator subgroups of free groups", Amer. J. Math. 77 (1955), 929-931.

Higman, Graham, "Finite groups having isomorphic images in every finite group of which they are homomorphic images", Quart. J. Math. Oxford (Ser. 2) 6 (1955), 250-254.

1958

Chen, K.T., Fox, R.H. and Lyndon, R.C., "Free differential Calculus IV: The quotient groups of the lower central series", Ann. of Math. 68 (1958), 81-95.

1960

Mal'cev, A.I., "On free solvable groups", Dokl. Akad. Nauk SSSR 130 (1960), 495-498 (Russian), [English transl. Soviet Math. Dokl. 1 (1960), 65-68].

1962

Gruenberg, Karl, "The residual nilpotence of certain presentations of finite groups", Arch. Math. 13 (1962), 408-417.

Jacobson, Nathan, "Lie Algebras", Interscience Publ. (1962), New York.

Neumann, B.H., "On a theorem of Auslander and Lyndon", Arch. Math. 13 (1962), 4-9.

1964

Šmel'kin, A.L., "Wreath products and varieties of groups", Dokl. Akad. Nauk SSSR 157 (1964), 1063-1065 (Russian).

1965

Bachmuth, S., "Automorphisms of free metabelian groups", Trans. Amer. Math. Soc. 118 (1965), 93-104.

Dunwoody, M.J., "On verbal subgroups of free groups", Arch. Math. 16 (1965), 153-157.

Serre, Jean-Pierre, "Lie Algebras and Lie Groups", Mathematics Lecture Notes Series (1965), Benjamin/Cummings Publishing Co., Massachusetts.

Šmel'kin, A.L., "Wreath products and varieties of groups", Izv. Akad. Nauk SSSR Ser. Mat. 29 (1965), 149-170 (Russian).

1966

Bachmuth, S. and Hughes, I., "Applications of a theorem of Magnus", Arch. Math. 17 (1966), 380-382.

Kargapolov, M.I. and Remeslennikov, V.N., "The conjugacy problem for free solvable groups", Algebra i Logika 5 (1966), 15-25 (Russian).

Magnus, Wilhelm; Karrass, Abraham and Solitar, Donald, "Combinatorial Group Theory", Interscience Publ. (1966), New York.

Matthews, Jane, "The conjugacy problem in wreath products and free metabelian groups", Trans. Amer. Math. Soc. 121 (1966), 329-339.

1967

Baumslag, G. and Gruenberg, K.W., "Some reflections on cohomological dimension and freeness", J. Algebra 6 (1967), 394-409.

Cohen, D.E., "On the laws of a metabelian variety", J. Algebra 5 (1967), 267-273.

1968

Enright, Dennis E., "Triangular matrices over group rings", Doctoral Thesis, New York University (1968).

Ojanguren, Manuel, "Freier Präsentierungen endlicher Gruppen und zugehörige Darstellungen", Math. Zeitchr. 106 (1968) 293-311.

Passi, I.B.S., "Dimension subgroups", J. Algebra 9 (1968), 152-182.

Quillen, Daniel G., "On the associated graded ring of a group ring", J. Algebra 10 (1968), 411-418.

1969

Dunwoody, M.J., "The Magnus embedding", J. London Math. Soc. Ser. II, 44 (1969), 115-117.

Gupta, C.K., "A faithful matrix representation of certain centre-by-metabelian groups", J. Austral. Math. Soc. 10 (1969), 451-464.

Hoare, A.H.M., "Group rings and lower central series", J. London Math. Soc. (2) 1 (1969), 37-40.

1970

Gruenberg, Karl W., "Cohomological Topics in Group Theory", Lecture Notes in Math. 143 (1970), Springer-Verlag.

Hartley, B., "The residual nilpotence of wreath products", Proc. London Math. Soc. (3) 20 (1970), 365-392.

Moran, S., "Dimension subgroups mod n", Proc. Cambridge Phil. Soc. 68 (1970), 579-582.

Remeslennikov, V.N. and Sokolov, V.G., "Some properties of the Magnus embedding", Algebra i Logika 9 (1970), 566-578.

1971

Baumslag, Gilbert, "Lecture Notes on Nilpotent Groups", Regional Conference Series #2 (1971), American Mathematical Society.

Dunwoody, M.J., "The hopficity of F/R'", Bull. London Math. Soc. 3 (1971), 18-20.

Smith, Martha, "Group algebras", J. Algebra 18 (1971), 477-499.

1972

Bachmann, F. and Gruenenfelder, L., "Homological methods and the third dimension subgroup", Comment. Math. Helv. 47 (1972), 526-531.

Hurley, T.C., "Representations of some relatively free groups in power series rings", Proc. London Math. Soc. (3) 24 (1972), 257-294.

Rips, E., "On the fourth integer dimension subgroup", Israel J. Math. 12 (1972), 342-346.

Romanovskii, N.S., "A freeness theorem for groups with a single defining relation in varieties of solvable and nilpotent groups of given classes", Math. Sbornik 89 (1972), 93-99.

Sandling, Robert, "The dimension subgroup problem", J. Algebra 21 (1972), 216-231.

Sarkisyan, R.A., "Conjugacy in free polynilpotent groups", Algebra i Logika 11 (1972), 694-710 (Russian).

1973

Gupta, Chander Kanta, "The free centre-by-metabelian groups", J. Austral. Math. Soc. 16 (1973), 294-299.

Hurley, T.C., "On a problem of Fox", Invent. Math. 21 (1973), 193-141.

Mital, J.N. and Passi, I.B.S., "Annihilators of relation modules", J. Austral. Math. Soc. 16 (1973), 228-233.

Remeslennikov, V.N., "An example of a group finitely defined in the variety with an unsolvable equality problem", Algebra i Logika 12 (1973), 577-602 (Russian).

Wehrfritz, B.A.F., "Infinite Linear Groups", Ergeb. Math. Grenzgeb. 76 (1973), Springer-Verlag.

1974

Birman, Joan S., "An inverse function theorem for free groups", Proc. Amer. Math. Soc. 41 (1974), 634-638.

Birman, Joan S., "Braids, Links and Mapping Class Groups", Annals of Mathematics Studies 82 (1974), Princeton University Press.

Gupta, C.K. and Gupta, N.D., "Power series and matrix representations of certain relatively free groups", Proc. Second Internat. Conf. Theory of Groups, Canberra (1973), In: Lecture Notes in Math. 372, 318-329 (1974), Springer-Verlag.

Lewin, Jacques, "A matrix representation for associative algebras, I", Trans. Amer. Math. Soc. 188 (1974), 293-308.

Losey, Gerald, "N-series and filterations of the augmentation ideal", Canad. J. Math. 26 (1974), 962-977.

Romanovskii, N.S., "Some algorithmic problems for solvable groups", Algebra i Logika 13 (1974), 26-34 (Russian).

1975

Bergman, George M. and Dicks, Warren, "On universal derivations", J. Algebra 36 (1975), 193-211.

Passi, I.B.S., "Annihilators of relation modules", J. Pure Appl. Algebra 6 (1975), 235-237.

Stallings, John R., "Quotients of powers of the augmentation ideal in a group ring", In: Knots, Groups and 3-Manifolds, Annals of Math. Studies 84 (1975), 101-118.

1976

Gupta, N.D. and Passi, I.B.S., "Some properties of Fox subgroups of free groups", J. Algebra 43 (1976), 198-211.

1977

Gupta, Narain, "Fox subgroups of free groups", J. Pure Appl. Algebra 11 (1977), 1-7.

Hurley, T.C., "Residual properties of groups determined by ideals", Proc. Royal Irish Academy 77 (1977), 97-104.

Kuz'min, Yu. V., "Free centre-by-metabelian groups, Lie algebras and D-groups", Izv. Akad. Nauk SSSR Ser. Mat. 41 (1977), 3-33. [English Trnasl. Math. USSR-Izv. 11 (1977), 1-30.]

Lichtman, A.I., "The residual nilpotence of the augmentation ideal and the residual nilpotence of some classes of groups", Israel J. Math. 26 (1977), 276-293.

Lyndon, Roger C. and Schupp, Paul E., "Combinatorial Group Theory", Ergebnisse Math. Grenzgeb. 89 (1977), Springer-Verlag.

Passman, Donald S., "The Algebraic Structure of Group Rings", Interscience Publ. (1977), New York.

Tahara, Ken-Ichi, "On the structure of $Q_3(G)$ and the fourth dimension subgroup", Japan J. Math. (N.S.) 3 (1977), 381-396.

Thomson, M., "On a theorem of C.K. Gupta", Research notes (1977) [unpublished].

1978

Gupta, Chander Kanta, "Subgroups of free groups induced by certain products of augmentation ideals", Communications in Algebra 6 (1978), 1231-1238.

Gupta, Chander Kanta and Gupta, Narain Datt, "Generalized Magnus embeddings and some applications", Math. Zeitschr. 160 (1978), 75-87.

Krasnikov, A.F., "Generators of the group F/[N,N]", Mat. Zametki 24 (1978), 167-173 (Russian), [English transl. Math. Notes 24 (1979), 591-594.]

Krasnikov, A.F., "Nilpotent subgroups of relatively free groups", Algebra i Logika 17 (1978), 389-401 (Russian).

Sehgal, Sudarshan K., "Topics in Group Rings", Marcel Dekker Inc. (1978), New York.

1979

Gupta, Narain D., Laffey, Thomas J. and Thomson, Michael W., "On the higher relation modules of a finite group", J. Algebra 59 (1979), 172-187.

Passi, Inder Bir S., "Group Rings and their Augmentation Ideals", Lecture Notes in Math. #715 (1979), Springer-Verlag.

Sjogren, J.A., "Dimension and lower central subgroups", J. Pure Appl. Algebra 14 (1979), 175-194.

Thomson, M.W., "Representations of certain verbal wreath products by matrices", Math. Zeitschr. 167 (1979), 239-257.

1980

Baumslag, G., Strebel, R. and Thomson, M.W., "On the multiplicator of $F/\gamma_c(R)$", J. Pure Appl. Algebra 6 (1980), 121-132.

1981

Gupta, C.K. and Gupta, N.D., "Magnus embeddings for groups and group rings", Houston J. Math. 7 (1981), 43-52.

Gupta, Narain, "A problem of R.H. Fox", Canad. Math. Bull. 24 (1981), 129-136.

Gupta, N.D. and Passi, I.B.S., "The Fox modules of a finite group", Indian J. Pure Appl. Math. 12 (1981), 430-439.

Tahara, Ken-Ichi, "The augmentation quotients of group rings and the fifth dimension subgroups", J. Algebra 71 (1981), 141-173.

1982

Chandler, Bruce and Magnus, Wilhelm, "The History of Combinatorial Group Theory: a case study in the history of ideas", Studies in the History of Math. and Phys. Science 9 (1982), Springer-Verlag.

Gupta, C.K., "On the conjugacy problem for F/R'", Proc. Amer. Math. Soc. 85 (1982), 149-153.

Gupta, Narain, "On the dimension subgroups of metabelian groups", J. Pure Appl. Algebra 24 (1982), 1-6.

Hartley, B., "Dimension and lower central subgroups - Sjogren's theorem revisited", Lecture Notes 9 (1982), National University Singapore.

Huppert, B. and Blackburn, N., "Finite Group II", Grundlehren der math. Wissenschaften 242 (1982), Springer-Verlag.

Kuz'min, Yu. V., "On elements of finite order in free groups of some varieties", Math. Sbornik 119 (1982), 119-131 (Russian).

1983

Gupta, Chander Kanta, "Subgroups induced by certain ideals of free group rings", Comm. Algebra 11 (1983), 2519-2525.

Gupta, Narain and Levin, Frank, "On the Lie ideals of a ring", J. Algebra 81 (1983), 225-231.

Passi, I.B.S. and Vermani, L.R., "Dimension subgroups and Schur multiplicator", J. Pure Appl. Algebra 30 (1983), 61-67.

1984

Gupta, C.K. and Passi, I.B.S., "Magnus embeddings and residual nilpotence", Preprint (1984), [J. Algebra 105 (1987), to appear].

Gupta, Narain, "Fox subgroups of free groups II", In: Contemporary

Mathematics 33 (1984), 223-231.

Gupta, Narain, "Sjogren's theorem for dimension subgroups - the metabelian case", Proc. Internat. Conf. on Combinatorial Group Theory and Topology, Utah (1984), In: Annals of Math. Study (to appear).

Gupta, N.D., Hales, A.W. and Passi, I.B.S., "Dimension subgroups of metabelian groups", J. reine u. angew. Math. 346 (1984), 194-198.

Hartley, B., "A note on free presentations and residual nilpotence", J. Pure Appl. Algebra 33 (1984), 31-39.

Hartley, B., "Topics in the theory of nilpotent groups", In: Group Theory essays for Philip Hall (1984), Academic Press.

Kuz'min, Yu. V., "The structure of free groups of certain varieties", Math. Sbornik 125 (1984), 128-142 (Russian), [English Transl. Math. USSR Sbornik 53 (1986), 131-145.]

Lang, Serge, "Algebra", Addison-Wesley Publishing Co. (1984), California.

Lyndon, Roger, "Problems in Combinatorial Group Theory", Proc. Internat. Conf. Combinatorial Group Theory and Topology, Utah (1984), In: Annals of Math. Study (to appear).

Passi, I.B.S., "The free group ring", In: Algebra and its Applications, Lecture Notes in Pure Appl. Math. 91 (1984), Marcel Dekk

Stöhr, Ralph, "On Gupta representations of central extensions", Math. Zeitschr. 187 (1984), 259-267.

Yunus, I.A., "On a problem of Fox", Soviet Math. Dokl. 30 (1984), 346-350.

Yunus, I.A., "Generalized Magnus embeddings and some residual properties of groups and group rings", Vestn. Mosk. Univ. Ser. 1, No. 3 (1984), 85-88 (Russian).

1985

Cliff, G. and Hartley, B., "Sjogren's theorem on dimension subgroups", Research Report 325 (1985), National University of Singapore [J. Pure Appl. Algebra (to appear)].

Gupta, Narain, "A theorem of Sjogren and Hartley on dimension subgroups", Preprint (1985), [Publ. Math. Debrecen (to appear)].

Gupta, Narain and Tahara, Ken-Ichi, "Dimension and lower central subgroups of metabelian p-groups", Nagoya Math. J. 100

(1985), 127-133.

Hales, Alfred W., "Stable augmentation quotients of abelian groups", Pacific J. Math. 118 (1985), 401-410.

Hartley, B., "A note on a lemma of Sjogren relating to dimension subgroups", Research Report 229 (1985), National University of Singapore.

Hurley, T.C., "identifications in a free group", Preprint (1985).

Khambadkone, Meera, "On the structure of augmentation ideals in group rings", J. Pure Appl. Algebra 35 (1985), 35-45.

Kuz'min, Yu. V., "Elements of order 2 in a free centre by solvable group", Mat. Zametki 37 (1985), 643-652 (Russian).

Röhl, Frank, "Review and some critical comments on a paper of Grün concerning the dimension subgroup conjecture", Pol. Soc. Pras. Mat. 16 (1985), 11-27.

Stöhr, Ralph, "On torsion in free central extensions of some torsion free groups", Preprint (1985), [J. Pure Appl. Algebra (to appear)].

1986

Gupta, Chander Kanta and Levin, Frank, "Dimension subgroups of free centre-by-metabelian groups", Ill. J. Math. 30 (1986), 258-273.

Gupta, C.K., Gupta, N.D. and Levin, F., "On dimension subgroups relative to certain product ideals", In: Group Theory Brixen 86, Springer Lecture Notes (to appear).

Gupta, Narain, "Dimension subgroups (Sjogren's Theorem) [Research Report (1986)].

Passi, I.B.S., Sucheta and Tahara, Ken-Ichi, "Dimension subgroups and Schur multiplicator - III", Preprint (1986) [Japan J. Math. (to appear)].

Stöhr, Ralph, "Fixed points on higher relation modules of finite groups", Preprint (1986), [J. Algebra (to appear)].

1987

Gupta, Narain, "Dimension subgroups of metabelian p-groups", [J. Pure Appl. Algebra (to appear)].

467